ENVIRONMENTAL AND HEALTH IMPACT ASSESSMENT OF DEVELOPMENT PROJECTS

A Handbook for Practitioners

Edited by
Robert G. H. Turnbull

ENVIRONMENTAL AND HEALTH IMPACT ASSESSMENT OF DEVELOPMENT PROJECTS

A Handbook for Practitioners

Published on behalf of the World Health Organization Regional Office for Europe and the Centre for Environmental Management and Planning

by

First published by
TAYLOR & FRANCIS LTD

Reprinted 2001 by Taylor & Francis Ltd
2 Park Square, Milton Park, Abingdon, Oxon, OX14 4RN

Taylor & Francis is an imprint of the Taylor & Francis Group

Transferred to Digital Printing 2006

© 1992 TAYLOR & FRANCIS LTD

British Library Cataloguing in Publication Data
Environmental and health impact assessment of
development projects.
333.71

ISBN 1851665978

Library of Congress Cataloguing-in-Publication Data
Environmental and health impact assessment of
development projects: a handbook for practitioners.
    p.   cm.
    Includes bibliographical references and index.
    ISBN 1-85166-597-8
    1. Industrial development projects –
Environmental aspects.   2. Industrial development
projects – Health aspects.   3. Rural development
projects – Environmental aspects.   4. Rural
development projects – Health aspects.   I. World
Health Organization. Regional Office for Europe.
II. University of Aberdeen. Center for
Environmental Management and Planning.
    [DNLM: 1. Developing Countries.   2.
Environmental Health.   3. Program Evaluation.   4.
Public Health.   5. Public Policy. WA 395 E61]
RA579.E59   1991
333.7′15′091724–dc20
DNLM/DLC
for Library of Congress   91–19873
                                                            CIP

The views expressed in this publication are those of the authors and do not necessarily represent the decisions or the stated policy of the World Health Organization.

No responsibility is assumed by the Publisher for any injury and/or damage to persons or property as a matter of products liability, negligence or otherwise, or from any use or operation of any methods, products, instructions or ideas contained in the material herein.

Publisher's Note
The publisher has gone to great lengths to ensure the quality of this reprint but points out that some imperfections in the original may be apparent

Contents

Chapter 8—CASE STUDY EXAMPLES

Foreword

The WHO strategy of health for all by the year 2000 urges the implementation of healthy development policies at both the national and local levels. Such policies should take into consideration not only possible adverse environmental effects but also possible human health consequences. Increasingly, countries are becoming aware that unless both environmental and human health aspects are carefully integrated into development projects, short-term economic advantages can turn into long-term health nightmares. Environmental health impact assessment offers an appropriate mechanism for maximizing the advantages of such projects while minimizing possible environmental and health damage.

Environmental impact assessment identifies, at an early stage, the impact on both the environment and human health that planned development projects may have. Once potential adverse effects are identified, the necessary safeguards can be introduced. Environmental health impact assessment thus puts into action the sound public health principle: prevention is better than cure. In doing so, its overall purpose is to create an environment conducive to achieving physical and social well-being.

While this volume will be useful to any professional involved in planning or implementing development projects, I believe that it will be particularly relevant to those who have had little experience in integrating human health considerations into such projects. In this regard, this book will be of special value in addressing the needs of the countries of central and eastern Europe.

J.E. ASVALL
Regional Director
WHO Regional Office for Europe

Preface

This handbook for practitioners has been edited by Robert G. H. Turnbull, Senior Consultant, CEMP. It is based on twenty-ninepapers covering a wide range of environmental impact assessment (EIA) and environmental health impact assessment (EHIA) topics and written by an international group of experts from developed and developing countries. The papers were prepared as part of the 1987 and 1988 International Training Seminars on Environmental Assessment and Management, sponsored by the World Health Organization and the United Nations Development Programme, and organized annually since 1980 by the Centre for Environmental Management and Planning, CEMP, Aberdeen University, Scotland.

Reviewed extracts from these papers, which comprise the eight chapters of this volume, were prepared by CEMP, with the assistance of the Environmental Health Impact Laboratory of the Italian Public Health Institute in Rome. The list of these papers is given in Annex E.

The handbook is aimed at practitioners involved in EIA and EHIA, who are experienced in environmental or public health, toxicology or ecotoxicology.

This volume would not have been possible without the support and hard work of the many authors. In addition, the advice and input of the following individuals are gratefully acknowledged: A.-R. Bucchi, A. B. Chisholm, B. D. Clark, M. Davies, F. Ewing, D. O. Harrop, L. Stewart, R. G. H. Turnbull and G. Zapponi.

E. GIROULT
Regional Adviser
Environment and Health Planning and Ecology
WHO Regional Office for Europe

Chapter 1

Introduction to EIA and Health Assessment

1.1 GENERAL BACKGROUND

The World Health Organization (WHO) is concerned with the health impact of economic development projects and policy. Most often, these health impacts are secondary consequences of other environmental effects. The worldwide development of environmental impact assessment (EIA) procedures has provided a way to include and assess the health impacts of the same development projects.

The combination of environmental and health impact assessment (EHIA) procedures should lead to the identification of alternative development policies or projects less detrimental to health and the environment, and/or the provision of mitigation measures to compensate for potential environmental health impacts.

Where environmental assessment procedures exist, the prime objective is to ensure that likely impacts on human and health safety are properly assessed. This is likely to involve the development of methods for taking account of matters pertaining to health and safety, as well as considerable effort in the establishment of collaboration between environmental and public health professionals. In the absence of environmental assessment procedures, the objective must be to get them developed as environmental and health impact assessment procedures.

The World Health Organization is interested in environmental impact assessment only where:

- It provides for adequate assessment of human health and safety impact of development policies and projects or of consumer products.

- It encourages intersectoral collaboration especially between environmentalists and public health professionals.
- It provides for public information and community participation in environmental health issues.

1.2 THE VALUE OF HEALTH IMPACT ASSESSMENT WITHIN THE EIA SYSTEM

The benefits of including health effects within EIA vary depending on the subject of the assessment and its physical and institutional setting. Not all environmental assessments need or ought to encompass health effects, but there is a strong argument that all initial scoping procedures should examine the possibility that the development or policy under review might have an impact on health.

The purpose of health impact assessment within EIA has been defined as 'the creation of an environment conducive to the achievement of physical and social well-being' (Giroult, 1985). The health component must therefore be a synthesis of a wide range of effects in relation to their impact on human welfare and is much broader than merely the prediction of disease or ill health resulting from a development project. A health expert should be best able to recognise these widely defined health effects.

The advantages accruing from strengthened health effects assessment within EIA would vary with local conditions. Where public health and other regulatory bodies do not have the resources or power to fully control the environmental health impacts of development decisions, or where air and water quality standards have not been established, health impact assessments within EIA would enable controls to be introduced on a case-by-case basis to prevent serious deterioration in health conditions. The health component of EIA also serves a useful purpose where there are established authorities and procedures to assess and control deleterious environmental health effects. Health effects of proposals are of particular interest to the public and their inclusion in an EIA ought to encourage realistic public debate and increase public confidence in the EIA process. It would also produce a more complete document for the decision-maker and may result in health effects receiving proper consideration in decisions to permit or modify proposals.

1.3 EVOLUTION OF ENVIRONMENTAL IMPACT ASSESSMENT

Since the 1950s, growing environmental awareness has increasingly focused attention on the interactions between development actions and their environmental consequences. In developed countries, this has led to the public demanding that environmental factors be explicitly considered in the decision-making process. A similar situation is now occurring in developing countries but in this case it is often the decision-makers in Environment and Health Departments of Government who are giving a lead.

Early attempts at project assessment were crude and often based upon technical feasibility studies and cost–benefit analysis (CBA). CBA was developed as a means of expressing all impacts in terms of resource costs valued in monetary terms. However, a number of major developers which were assessed using CBA techniques, caused considerable public disquiet. Flaws in CBA became more apparent and one consequence was the development of a new evaluation approach which came to be known as environmental impact assessment (EIA). The concept of EIA was also seen as a potentially useful tool to assist environmental lobbies and their cause. It has evolved as a comprehensive approach to evaluation, in which environmental considerations, as well as economic and technical considerations, are given their proper weight in the decision-making process.

When EIA was first conceived, it was regarded as an 'add on' component to CBA, and was designed to incorporate all those potential impacts that had proved troublesome in CBA (Council of Environmental Quality, 1978). As a consequence, early practitioners used the system as a means of collecting information but often failed to comprehend the policy environment in which the development was proposed. More fundamental questions are now being posed such as:

(i) Is the development required?
(ii) What are the relevant alternatives which would provide the same benefits and how do respective EIAs compare?
(iii) What is the appropriate level of public safety in relation to hazardous technologies?
(iv) What degree of environmental protection should be guaranteed for areas of significant ecological and landscape value?
(v) What are the short- and long-term health consequences?

EIA can, therefore, be thought of as a basic tool for the sound assessment of development proposals. Whilst EIA can play an important role in formulating environmentally sound policies and plans, and in their evaluation, this handbook for practitioners concentrates on the application of EIA at a project level.

1.4 THE RATIONALE FOR EIA (DEFINITIONS AND OBJECTIVES)

Until recently, projects were formulated and assessed according to technical, economic and political criteria. Potential environmental health and social impacts of projects were rarely considered, and then, only in the form of cost–benefit analysis which crudely attempted to place a monetary value upon non-economic variables such as the destruction of marine ecosystems or the social and health impacts of air pollution. Such restricted assessments of development projects resulted in unforeseen harmful impacts which reduced predicted benefits, through deleterious secondary effects.

The purpose of an EIA is to determine the potential environmental, social and health effects of a proposed development. It attempts to define and assess the physical, biological and socio-economic effects in a form that permits a logical and rational decision to be made. Attempts can be made to reduce potential adverse impacts through the identification of possible alternative sites and/or processes. There is no general and universally accepted definition of EIA. The great diversity of EIA definitions is illustrated by the following examples:

(i)   '...an activity designed to identify and predict the impact on the biogeophysical environment and on man's health and well-being of legislative proposals, policies, programmes, projects and operational procedures, and to interpret and communicate information about the impacts'.

(ii) '...to identify, predict and to describe in appropriate terms the pros and cons (penalties and benefits) of a proposed development. To be useful, the assessment needs to be communicated in terms understandable by the community and decision-makers and the pros and cons should be identified on the basis of criteria relevant to the countries affected'.

(iii) '...an assessment of all relevant environmental and resulting social effects which would result from a project'.

(iv) '...assessment consists in establishing quantitative values for selected parameters which indicate the quality of the environment before, during and after the action'.

(v) '... the systematic examination of the environmental consequences of projects, policies, plans and programmes. Its main aim is to provide decision-makers with an account of the implications of alternative courses of action before a decision is made'.

The above definitions provide a broad indication of the objectives of EIA, but illustrate differing concepts of EIA. In this handbook for practitioners, EIA is considered to be a technical exercise, the object of which is to provide decision-makers and the public with an account of the implications of proposed courses of action before a decision is taken.

The results of the assessment are assembled into a document known as an environmental impact statement (EIS) which sets out the beneficial and adverse impacts considered to be relevant to the project, plan or policy under consideration. The completed EIS is one component of the information upon which the decision-maker makes a choice. For example, other factors such as unemployment, energy requirements or national policies may influence the outcome of the final decision.

The likely consequences of adopting a particular course of action, may be reduced by introducing appropriate monitoring and auditing programmes for any deleterious impacts that may have been identified in the assessment process.

1.5 THE ADVANTAGES OF USING EIA

EIA can assist in the efficient use of valuable natural and human resources by developers and decision-makers. It can reduce costs and the time taken to reach a decision by ensuring that subjectivity and duplication of effort are eliminated, and by identifying and quantifying, primary and secondary consequences, which may require the introduction of expensive pollution control equipment, compensation or other costs at a later date.

EIA can improve the efficiency of decision-making, but to be effective it should be implemented at the project planning and design stage. It must be an integral component in the design of a project, rather than something utilised after the design phase is complete. Preferably, EIAs should be part of the decision-making process which has a number of decision points in the project planning cycle. This means a continuous feedback between EIA findings, project design and locations. They can be used at an early stage to test alternative project designs, and in the choice of the project design with the greatest benefits and minimum harmful effects. EIA, therefore, can be used to investigate and avoid harmful impacts and to increase likely benefits.

The emergence of an optimum alternative in terms of the objectives relevant to a proposed project means that EIAs may have significant long-term financial advantages. The identification of a potential problem early in project planning may result in considerable financial and time savings. For example, the abandonment of a project would save capital costs if all alternative designs or locations proved unsuitable in terms of likely detrimental effects. More likely, however, design modifcations may reduce the need for expensive ameliorating action once a project becomes operational. A development which is not assessed for its likely impacts, may cause serious social and health problems. For example, a proposed dam and reservoir may have health effects which may require an expensive health care programme. The wrong locations for settling the displaced population may result in agriculture failure and the need for food supplies to be sent to the relocated people from other areas.

1.6 THE POSSIBLE DISADVANTAGES OF EIA

EIA cannot be a cure for all environmental ills. In certain areas of decision-making, its application may be restricted because of a number of difficulties. These include:

- National policies where assessment ought to be confined to consideration of the implications of such policies.
- Proposals arising from general 'need' such as provision of employment, where it may be appropriate to use EIA as a means of justification.
- When disputes arise, its effectiveness and impartiality may be impaired as a developer may already be fully committed to the specific project.
- The provision of inadequate time for the assessment and modification to the proposals because of public controversy.
- Bias in favour of the agency responsible for the assessment. This may take the form of a comprehensive public relations document rather than an aid to decision-making. An encyclopedic type of EIA may be produced to persuade the authorising agency that the development should proceed.
- Superficiality in describing important key issues which relate to the development, and lack of information on the specificity of impacts, such as the identification of the important parameters of an impact; the timing and duration of the impacts in the construction or operational phases.

The advantages of an EIA system may be reduced if the procedures adopted do not help to overcome some of the above issues.

1.7 THE SCOPE OF EIA

In principle, EIA should apply to all actions likely to have a significant environmental effect. The potential scope of a comprehensive EIA system is considerable and could include the tiered appraisal of policies, plans, programmes and development projects.

The potential advantages of a tiered approach over a procedure which is restricted to development projects are as follows:

(i)   At the development project stage, the available options are often severely limited by earlier decisions made at a higher level. Misspecification of a project assessment (e.g. a road scheme) may occur if the higher levels (e.g. transport policies) were not subject to such evaluation. The issue may be where environmental impacts will occur, and not whether they ought to.

(ii) Assessment of individual projects can only be conducted once proposals have been made. It cannot guarantee optimum site selection, and a thorough assessment of all alternative actions may be expensive and time-consuming.

(iii) The scope of viable alternatives decreases at the project level; and the willingness to contemplate alternatives.

(iv) The available time for collection and analysis of environmental data becomes increasingly restricted at the lower tier unless a programme of establishing environmental base-line data is undertaken independently of individual project EIAs.

(v)   When projects are individually small in size, but collectively large in number (e.g. housing), an EIA at the plan and programme stages may lead to a reduction in the time required for evaluation.

When plans are based on a sound environmental assessment, specific non-conforming project proposals are likely to require detailed environmental assessment. The preparation of plans must be based on adequate data relating to the existing environment and the implications of possible changes. The systematic collection, analysis, storage and regular updating of such data must improve the quality of subsequent EIAs and reduce time and costs. EIA and land use planning must be complementary to each other.

1.8 SELECTION OF PROJECTS FOR EIA

The selection of projects which should be subject to EIA has resulted in a build-up of case law which indicates those development projects for which an authorising agency should request an EIA. In countries without such procedures, no such guidelines exist, the decision being dependent upon the scale of the project, its environmental setting, uniqueness of the project and the likely degree of public opposition. Due to different approaches disparities exist as to the number of EIAs prepared.

Project selection for EIA can be undertaken in a number of ways. Development types with specific criteria and thresholds of size, cost and/or power requirements, may automatically require an EIA. In some countries, EIAs are mandatory for certain classes of development which exceed a specified financial threshold. This, however, neglects the importance of the environmental setting of a project. The extent and significance of a particular impact depends not only on the causative agent, such as the amount of a pollutant, but also on the sensitivity of the receiving environment. Alternatively, environmentally sensitive areas can be designated in which an EIA is required for specific development projects. This approach implies that only specified developments are detrimental to certain environmental features.

The most suitable approach involves consideration of both aspects. Such a procedure could set out a list of development project types requiring an EIA and those which did not require an EIA. A screening elimination process could then be applied to identify developments on either list which require an EIA. An initial assessment of the project can determine the need or otherwise of a detailed EIA. A preliminary assessment would involve examination of the following components:

(i)   An environmental description of possible sites where physical, socio-economic and human health may be directly or indirectly affected by the project. This provides the base-line against which the potential impacts can be assessed. Existing land uses can be identified including those which are in conflict or which are complementary to the project.

(ii) The potential environmental effects can be identified with an indication of their magnitude in relation to the prevailing conditions. The assessment can identify those interests which may be either beneficially or detrimentally affected by the project.

(iii) A description of the possible mechanisms of environmental control as well as a list of relevant environmental standards can be prepared.

1.9 RESPONSIBILITY FOR IMPLEMENTATION OF EIAs

The responsibility for undertaking an EIA depends upon the particular EIA system planned or operating and the nature of the project. Four major alternatives can be identified as follows:

(i) The authorising agency assumes responsibility although the expertise, time and financial implications may offset any potential advantages of impartiality.
(ii) The developer undertakes or commissions consultants to undertake the EIA. Difficulties relate to the impartiality of the assessment and the initial objectives of the proponent in undertaking the environmental assessment.
(iii) Shared responsibility between the authorising agency and the developer. The responsibilities of the authorising agency are developing the terms of reference and guiding the EIA, whilst the developer is responsible for the EIA costs and the actual assessment, often through the use of consultants.
(iv) An independent specialist body which is funded wholly or in part by government and which is able to call upon information from all parties undertakes the EIA. Such an agency should be totally impartial in the consideration of both private and public development proposals.

Impartiality may be improved in the following ways:

(i) Preparing generic or specific guidelines or minimum standards for the form and content of an EIA.
(ii) Supervision by a reviewing or controlling body with no vested interest in the project.
(iii) Mandatory consultation with relevant and competent organisations.
(iv) Publication and provision for public discussion of the impact statements.

1.10 REVIEW OF EIAs

The reviewing authority responsible for ensuring that the EIA was undertaken in accordance with the terms of reference may be the authority from which authorisation for the development project is requested, or an independent agency. Questions of impartiality may arise when the authorising agency has been responsible for the EIA. Any suspicion of bias may be removed by the use of an independent review agency.

The function of the Review Authority may include:

(i) Defining the 'scope' of the assessment, i.e. which projects should be subjected to a full or partial EIA.

(ii) Producing general or specific guidelines and advice on methods of EIA.

(iii) Formulating the terms of reference and initiate a detailed EIA.

(iv) Ensuring that the EIA had been adequately completed within the terms of reference.

(v) Submitting the EIS together with any separate contributions from other organisations, with recommendations to the appropriate authorising agency.

(vi) Acting as a focus for the exchange of information and opinions concerning environmental affairs.

The EIA should not be regarded as a procedure which is only used at the decision-making stages. It can be adapted after the decision, to ensure that the project conforms to the standards detailed in the relevant permissions, to provide a data base for any subsequent impact study, and to allow the monitoring and control programme to adapt to changing circumstances or increased knowledge.

1.11 PUBLIC INFORMATION AND PARTICIPATION IN EIA

Public involvement is an integral part of any EIA system. Efforts should be made to obtain the views of, and to inform, the public and other interest groups who may be directly or indirectly affected by a development project. The authorising agency may not always identify the environmental issues which the public perceives to be important and they may also lack the detailed local knowledge that the public possesses.

Advantages of participation may lead to the following:

- The provision of information about the local environmental, economic and social systems.
- The possible identification of alternative actions.
- An increase in the acceptability of the project as the public will better understand the reasons for the project.
- Minimising conflict and delay.

Problems may nevertheless arise. Public participation, in the short-term, may be time-consuming, increase costs, and participants may be unrepresentative of the community affected. In spite of these problems, many countries are actively encouraging public involvement in EIA.

1.12 FINANCIAL ASPECTS OF EIA

Initially, EIAs may be expensive to implement, particularly in areas where little is known about existing environmental and social conditions. Design changes produced as a result of EIA findings may also increase capital costs. It can be argued, however, that the avoidance of deleterious impacts and the maximisation of beneficial impacts will outweigh the costs of an EIA system in the long-term. The cost of an EIA system should decline when processes and techniques have been established and when assessment personnel become accustomed to their tasks.

The investigation of impacts at an early stage of project planning may well save money by speeding up the process of implementing a proposal. The costs of EIA are commensurate with the complexity and significance of the problem and the level of detailed investigation required. Costs may be borne by the proponent of the development project, or by the authorising authority.

The overall benefits of EIA have not been determined due to the difficulty in assigning monetary values to such benefits. Many environmental amenities that would otherwise have been degraded or destroyed have a unique value, and over time will far outweigh EIA costs. The use of EIA has allowed the choice of options which are both environmentally and economically superior to the original choices available.

1.13 SUMMARY

There is a general recognition that environmental consideration ought to be integrated into the planning of a decision-making framework, but differences exist as to the exact form that such integration should take. The administrative structures within which the EIA process operates also vary. EIA procedures have been implemented through legislative and administrative regulations, and/or by integration with planning or other authorisation systems.

During the evaluation of development proposals, EIA can ensure that environmental aspects are given equal status with economic, technical and social considerations. Attention can also be given to immediate impacts, and indirect, secondary and long-term effects. It is necessary to stress the importance of a proper framework for deciding which project development activities ought to be subject to EIA, and the limitations placed on project EIAs by decision made at policy or plan-making levels.

REFERENCES AND BIBLIOGRAPHY

Council of Environmental Quality (1978) National Environmental Policy Act — Regulations, Proposed Implementation of Procedural Provisions, US Executive Office of the President Washington, *Federal Register*, **43**(112), 25230–5.

Council of the European Community (1985) *Council Directive on the Assessment of the Effects of Certain Public and Private Projects on the Environment* (6553/85 Env 92).

Environmental Resources Ltd (1983) *Environmental Health Impact Assessment of Irrigated Agricultural Development Projects. Guidelines and Recommendations*, WHO Regional Office for Europe, Copenhagen.

Environmental Resources Ltd (1985) *Environmental Health Impact Assessment of Urban Development Projects, Guidelines and Recommendations*, WHO Regional Office for Europe, Copenhagen.

Fairbank, W., Donaldson, R.J. and Radcliffe, J.W. (1972) *British Steel Corporation Proposed Complex — Redcar Phases II and III.*

Giroult, E. (1985) *The Health Component of Environmental Impact Assessment*, paper presented at International Seminar on Environmental Impact Assessment, July 1985, Aberdeen, Scotland.

World Health Organization (1986) *Working Group on the Health and Safety Component of EIA Summary Report*, Regional Office for Europe, Copenhagen ICP/RUD 004/m/(S).

Chapter 2

The Nature of Environmental Impact Assessment

2.1 INTRODUCTION

Administrative procedures for environmental impact assessment vary from country to country. There is no single procedure which is common to all countries. Despite this, however, it is possible to identify specific characteristics and aspects of the EIA process. Before considering the process, it is useful to list the main objectives of EIA. They are:

- To identify beneficial and adverse environmental impacts.
- To suggest mitigation actions which might reduce or prevent adverse impacts.
- To identify and describe the residual adverse impacts which cannot be mitigated.
- To identify appropriate monitoring strategies to 'track' impacts and provide an 'early warning' system
- To incorporate environmental information into the decision-making process relating to development projects.
- To aid selection of the 'optimum' alternative, here alternative sites or project designs are being investigated in an EIA study.

Over the years, considerable attention has focused on the variety of institutional procedures which exist for the organisation of national EIA systems. These procedures are generally administrative in nature and provide for the production of environmental impact statements (EISs) to a uniform standard in accordance with established rules. However, procedures by themselves cannot ensure that EISs will contain structured information, produced in a 'scientific' manner in the best interests of the decision-making process. For this reason, methods and techniques are

13

required to aid identification and the assessment of impacts, and to ensure that the best possible information is made available to decision-makers and the public. As all methods involve impacts, there is a need to consider the nature of impacts as they are the central concern of EIA methods and techniques.

2.2 THE POLITICAL DIMENSION OF IMPACTS

The meaning and nature of the impact concept is important in that it may influence the nature and operational characteristics of EIA methods, and the final content of EISs. An impact is an event which is a direct consequence of a prior event. Within EIA procedures there is a constant selection process in operation to decide which events should or should not be included in the EIS. For example, the use of mechanical equipment during site preparation will give rise to increased ambient noise levels which are a direct result of the development activity. However, it is the indirect or secondary-order 'impact' of increased noise levels that are likely to disrupt the sleep of local residents at night, or during the day by making normal conversation more difficult; they would be included as impacts.

A second-order event may result in further events, some of which may not be perceived by those experiencing them. Individuals may blame increased noise levels for disrupted sleep, but not realise that increased family tension is the cause of this factor. For the development operators, the impacts of increased noise levels may be thought of only in terms of the number of complaints received. This example illustrates some of the complexities involved in the concept of impact.

The social or political choices involved in EIA work are not made by society as a whole. Societies are made up of groups and individuals with differing interests and values, consequently, the allocation of significance to environmental changes depends upon those who are aware of the change involved. Another aspect of this social aspect of 'impacts' is the time element. As techniques develop and more is understood of the relationships between environmental, social and health factors and their combined influence on individual well-being, 'new' impacts may be recognised. It follows, therefore, that improved knowledge of the casual relationships between development activity and environmental/social change leads to public recognition of impacts which were previously unsuspected.

The socio-political dimension of EIA is not new but the significance of this aspect of EIA is often ignored. For example, many EISs describe in considerable detail the likely concentrations of differing air or water pollutants at various locations relative to the source of emission, but fail to relate these environmental changes, which are, in themselves, neither good nor bad, to the interests and concerns of those groups which may use the resource being affected. In other words, environmental changes are not converted into impacts and described in terms of their consequences for people. This, of course, cannot be done for all changes such as ecological changes which may affect specialised social groups such as scientists, and may only be considered impacts by such groups. An attempt, however, should be made to trace all changes to social interests and how the population will be affected.

Tracing environmental changes to people is important in that beneficial or harmful impacts from projects are not evenly and homogeneously distributed among social groups or individuals. Everyone does not benefit or suffer equally, consequently, it is important for decision-makers and the public to appreciate that certain social groups may be subject to more harmful and fewer beneficial impacts, while other groups may be in a more favourable position. This type of information is essential for sensible public debate on the future of a proposed development project.

In addition to being a technical exercise, EIA involves value judgements throughout the process. EIA is also an activity in which 'objective' scientific description and prediction is a crucial component. For example, the prediction of water pollution impacts requires the use of dispersion models which provide data on concentrations at varying locations. Scientific method and a balanced viewpoint are important, but the role of value judgements throughout the EIA process has tended to be ignored. Politics, as well as science, are part of the EIA process. This duality needs to be appreciated by all those involved in EIA procedures.

2.3 IMPACT CHARACTERISTICS

Initial impacts on one component in an environmental system can have repercussions for others which may be nearby or distant from the component immediately affected. Depending upon the structure and functioning of the particular environmental system being stressed by a development, an initial impact can result in further impacts. When describing and discussing impacts on an EIS, therefore, it is useful to consider the range of characteristics which impacts exhibit. They are as follows:

Spatial dimension: Impacts can occur in the immediate vicinity of a project — for example, fluorosis caused by an aluminium smelter. Alternatively, they can occur at considerable distances from an installation. For example, hydrocarbon combustion in power plants in northern central Europe is considered to be a major cause of surface water acidification and tree deaths in Scandinavia.

Time dimension: Some impacts can occur immediately, for example noise effects; others are not apparent until a considerable time period has elapsed, for example an ecosystem may be able to absorb wastes emitted from an installation for many years until, finally, a critical threshold is reached and a change in the system is observed.

Reversibility: Some impacts are irreversible; others can be reversed either naturally or artificially.

Probability: As impact predictions refer to future effects there is a level of uncertainty associated with them. Each impact has a likelihood of occurrence. Generally, it is not possible to be deterministic and be certain regarding the occurrence of an impact and its likely scale. The topic will be discussed in detail later.

Beneficial/adverse: Some impacts are beneficial whereas others are adverse or harmful. It must not be assumed that all impacts are adverse.

Environmental social distribution: This characteristic relates, in part, to the spatial dimension of impacts. Certain species, ecosystems or social groups may be affected by more than one impact. In fact they may be affected by multiple impacts, for example noise, air pollutants and odour. They may be subject to a mix of adverse and beneficial consequences. This is especially the case if social groups are concerned. Some groups may gain employment at higher salary levels than previously and suffer no adverse impacts. Other groups may suffer only adverse impacts and gain no benefits. Generally, the 'real situation' will lie between the two extremes. Therefore, the overall distribution of impacts and their cumulative consequences are an important aspect of EIA studies.

2.4 THE EIA PROCESS IN RELATION TO THE PROJECT PLANNING CYCLE

To be effective an environmental impact assessment (EIA) procedure need only be applied to those actions which are considered to cause significant environmental consequences. It is, consequently, important to establish mechanisms for the selection of actions requiring EIA. Such a process of selection is termed *screening*. The next stage in EIA is determining which issues should be examined in the EIA, this activity is often termed *scoping*. In reality screening and scoping activities overlap, in that some methods not only provide a screening function, but also allow the identification of issues which require detailed examination. In addition, methods developed to assist the identification of potential impacts (e.g. impact matrices) may also be employed in impact assessment activities.

This section sets out the main elements of the EIA process in relation to the project planning cycle, and discusses the activities which may be implemented at each stage. Reference is made to a number of screening and scoping activities and to their strength and weaknesses. No recommendation is made to any particular approach to screening and scoping.

2.4.1 Pre-Feasibilty Study

Detailed designs of the development project are unlikely to be available at this stage, but the basic nature of the project together with some basic facts about the project and site or sites may be known. At this stage, the project may be subject to *screening* to decide whether a full comprehensive EIS must be prepared.

2.4.2 Methods for Screening

A number of methods are available to assist screening. They include project thresholds, sensitive area criteria; positive and negative lists; matrices; and initial environmental evaluations (Annex A). Projects may be categorised by a number of parameters (Table 2.1), the values of which will be dependent upon the project. Particular values can then be selected to determine whether an EIA might be required. For example, projects with a capital cost greater than say £5 million could then be subject to an EIA. Project parameters, by one means or another, are incorporated in the various screening methods. An alternative two stage screening process in which an initial screening can be used to quickly identify those projects which do and do not require an EIA. A secondary screening can then be used, subsequently, to determine whether EIAs are needed for remaining projects (see Fig. 2.1).

Table 2.1
Key project parameters

The Project
Project type
Plant specification, equipment, layout, operation etc.
Size of project
  ● Output volume
  ● Raw materials consumed
  ● Energy requirements
  ● Labour force
  ● Capital cost
  ● Physical size

Project Location
  ● By land use category industrial estate, residential,
    agricultural, etc.
  ● By geographical criteria such as coast, mountain, etc.
  ● By carrying capacity including air and water pollution
    dispersion characteristics

Project Phasing
  ● Site preparation
  ● Construction
  ● Operation
  ● Modification
  ● Induced development
  ● Decommissioning

2.4.3 Overview on Screening Methods

Each of the methods has different strengths and weaknesses, consequently the combined use of such methods can overcome individual weaknesses. In general, EIA screening activities utilise positive and negative lists which may be supplemented with other criteria and questionnaires. For example, screening activities in Malaysia include sensitive area criteria and a screening questionnaire (Fig. 2.2). An examination is first made of the positive list in order to identify those projects requiring an EIA. Then the negative list is examined. Projects on this list are further screened to ensure that they are not situated in an environmentally sensitive area, but if they are located in a sensitive area then an EIA would be necessary. Those projects not on either list are then subject to the questionnaire which comprises six main areas of inquiry, namely:

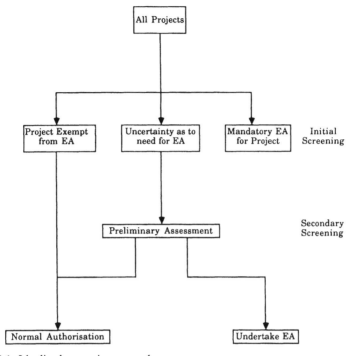

Fig. 2.1. Idealised screening procedure.

(i)   siting of project;
(ii) resource demand;
(iii) waste production;
(iv) labour requirements;
(v)   infrastructure needs;
(vi) regulations, guidelines and codes of practice.

Within each subject area the answers to a detailed series of questions then indicates the need for an EIA.

Associated with the different screening methods are varying degrees of simplicity and ease of operation. The application of a rigorous screening method to all projects could lead to delays to the system, consequently a gradual increase in complexity, in accordance with the difficulty in determining whether an EIA is needed for an individual project, may prove a beneficial approach.

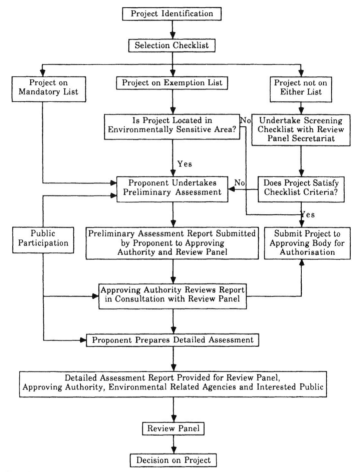

Fig. 2.2. Screening method.

2.4.4 Initial Environmental Evaluation

Some countries consider these approaches to be too deterministic and have initiated a more flexible case-by-case approach determined by an *initial environmental evaluation* (IEE). Basically it is a 'mini-EIA'. Even without detailed project information it is possible to identify likely significant impacts with enough certainty to be able to decide whether a full EIA should be prepared. Mistakes, of course, are likely but they are considered to be insufficiently important to outweigh the benefits of an overall approach which, as its major objective, attempts to ensure that manpower and financial resources are expended in assessing only those projects which need careful appraisal. In an initial environmental evaluation, information on actual impacts of similar projects, data on emission factors (often available from central government departments or international organisations) and basic knowledge of the site and surrounds, are brought together to form an opinion as to whether a full EIA is required.

2.4.5 Feasibility Study

At this stage, EIA work should be carried out in conjunction with economic, technical and design work. If the EIA can begin early then its contribution will be more effective at all stages of the project development. If EIA is treated as an 'add-on' once the design process has been finalised, and major environmental problems are identified then it may be time-consuming and expensive to integrate mitigation measures. EIA functions better as a 'preventative measure' rather than a 'cure'.

It is during this stage in the project cycle that most of the EIA process occurs. Once a decision has been made to prepare an EIS, it is necessary to identify the main impacts to be investigated in detail, the aim being to focus the EIA work only on those impacts of significance rather than take an all-embracing approach and study all possible impacts. In this way, the preparation of an EIS can be made cost-effective. The process of deciding on the impacts to be investigated is known as *scoping* and is becoming an increasingly important initial activity in any EIA.

2.4.6 Scoping

Scoping is the term given to the process of developing and selecting alternatives to a proposed action and identifying the issues to be considered in an EIA. Essentially, it is a procedure designed to establish the terms of reference for an EIA. Its aims are to:

- identify concerns and issues requiring consideration;
- facilitate an efficient EIS preparation process;
- enable those responsible for EIA to brief the study team on the alternatives and impacts to be considered at different depths of analysis;
- provide an opportunity for public involvement;
- save time.

Scoping is not a discrete exercise as it may continue into the planning and design phase depending upon new issues that may arise for consideration. Equally, some of the activities under screening may serve the same function as scoping. For example, the use of the IEE or matrices approach.

Scoping requires a number of procedural aspects to be considered. They include:

- responsibility for scoping;
- how is the scoping of an EIA defined in terms of the identification and selection of alternatives and the determination of significant issues;
- how are interested parties and the public to be involved in scoping.

2.4.6.1 Responsibility for scoping

Responsibility for scoping may rest with the authorising agency, where guidance is given to the project proponent on what issues should be examined — the terms of reference (TOR) for the EIA. In other cases scoping may be implemented by the proponent. In situations where there exists limited exchange of information and concerns between the proponent, relevant agencies and the public, there is the possibility that significant issues may be overlooked. This may then give rise to the rejection of the EIS or delay to the project while omitted issues are investigated.

2.4.6.2 Identification and selection of alternatives

The consideration of alternatives to the proposed project is one of the key aspects of EIA, in that it provides a means by which project assumptions, goals and needs can be examined. Consideration of alternatives provides for the examination of different mechanisms to achieve a stated objective and assists the decision-makers in the choice of an alternative which has the least adverse and greatest beneficial environmental social and economic consequences. While this concept may be readily accepted, problems invariably occur when consideration is given to the following aspects:

(i)   What is a reasonable range of alternatives to be considered either in terms of number or variety?
(ii) How should alternatives be identified?
(iii) What level of examination should be applied to each alternative?

2.4.6.3 Range of alternatives

Ensuring that a range of alternatives are examined is important, since failure to do so may result in the EIS being challenged on the basis of viable alternatives being omitted. This issue is often addressed by requiring that only 'practical' rather than all 'feasible' alternatives are considered.

2.4.6.4 Public involvement in scoping

Each country evolves its own approach to scoping depending upon the abilities and willingness of all parties to become involved in the exercise. Many papers have been written on the wider subject of public involvement in EIA and the mechanisms which may be adopted.

Given the extensive literature concerning public involvement which exists, the only approach described is that of the Canadian Federal Environmental Assessment Panels. The Panels, which comprise four to six experts, are established to examine the environmental and related implications of a particular project. The Panel is responsible for issuing guidelines for preparing an EIS and reviewing the completed EIS. The Panel is responsible for the scoping activity in that it identifies the issues which must be considered in the EIS. To aid this task, Panels will often request public views by written comment, by workshops or public meetings, before completing the preparation of the guidelines (FEARO, 1980). These public meetings are structured to allow:

(i)   the Panel to receive the views and questions of the local people;
(ii) the project proponents to have an opportunity to reply to questions and comments;
(iii) the agencies having an interest in the project to present their views concerning the project;
(iv) contentious issues to be discussed in order that the Panel may take account of all views on the subject.

Great care is required in the mechanisms adopted for seeking public involvement and it is important that the process is reinforced by feedback to the public of the results of their expressions of concern. In the Canadian context this is achieved by publishing the guidelines on EIS. The EIS is published for comment and if necessary a statement of any deficiencies in the EIS is also made public.

2.4.6.5 Overview on scoping

Both screening and scoping activities are important stages in EIAs, although their exact boundaries are often blurred. A variety of approaches are available each with various strengths and weaknesses. It is, therefore, important that appropriate methods, perhaps in combination, relevant to the needs of individual countries are adopted.

While approaches to screening are reasonably well documented, scoping does not have any specific methods, and hence only a few documents focusing upon this subject exist. Those which do tend to report the

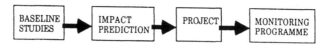

Fig. 2.3. Relationship of baseline studies to impact prediction.

advantages which can be gained from scoping. In an efficient scoping procedure, the unnecessary expenditure of time and financial resources on irrelevant issues are minimised, essentially streamlining the EIA. With the aid of involvement of the public, acceptable terms of reference for the EIA can be developed thereby reducing the likelihood of a major controversy once an EIS has been prepared. The scoping activities should also assist in the coordination of an action from the various agencies involved in the assessment of a proposed project. This is, however, the theory; it is equally possible that early public involvement and inter-agency politics could cause delays to the EIA rather than minimise them. It is not possible to suggest which view may be correct, since little evidence exists for either at present. As scoping relies to a great extent upon the exchange of information and concerns between the interested parties, including the public and appropriate organisations is essential if difficulties are to be avoided.

2.4.6.6 Impact identification
This activity occurs either as a part of formal scoping, whether initiated in a limited or extensive manner, or at the beginning of any EIA study. To identify impacts requires the bringing together of project data and information on the environment likely to be affected.

Until recently, this part of the EIA process was considered to follow the steps indicated in Fig. 2.3. Project data were obtained from the proponent and data on existing environmental conditions, often known as base-line data, obtained from field work or from written records. Once these two types of data had been obtained they could be brought together and likely impacts identified and predicted. After several years experience of EIA, especially in the United States and Canada, it was realised that base-line data collection was generally unfocussed and that data were collected which were not used in the actual EIA study. However, since the data had been obtained as part of the EIA, they were often included in the EIS and consequently, contributed to the size, length and costs of EIS production, which did not prove to be attractive, either to the decision-makers or members of the public.

A new strategy to overcome this problem has evolved in recent years. Basically, obtaining base-line data is now considered to be an evolving, expanding activity which continues until the production of the EIS. It is not a discrete action which only occurs at the beginning of EIA work. It certainly must occur otherwise it is not possible to begin the identification of likely impacts. However, once these have been identified only the requisite base-line data which enables these impacts to be investigated in detail are obtained. The collection of base-line data is focused by the needs of the EIA work on the impact prediction. As prediction and, later, impact evaluation proceeds, it may be that new base-line data needs are identified. If so, they should be collected. This strategy is an important change in the practice of EIA not yet widely adopted. However, if it can be seen to contribute to more cost-effective and shorter EISs, the wider adoption of this strategy is more likely to occur.

Having considered the relationship between base-line studies and impact identification it is important to consider EIA methods which have been developed to help with impact identification. The most well known are checklists, matrices and networks.

2.4.6.7 Identification of mitigating measures and monitoring schemes

Once impacts have been predicted and their likelihood of occurrence determined (if possible), some may be found to be both significant and adverse. At this point consideration should be given to the possibility of identifying and recommending mitigation schemes which might reduce or prevent the expected adverse impacts. These measures might include installation of additional pollution control equipment, landscaping to reduce visual intrusion or reduce noise transmission and process design changes. An assessment should be implemented to determine the extent to which each mitigating measure might reduce the impacts, or prevent them altogether. After this study of mitigation it is possible to include in the EIS only those impacts which are considered insufficiently serious to require mitigation, beneficial impacts (sometimes these can be enhanced) and those residual, significant adverse impacts which cannot be mitigated.

For the latter impacts it may be desirable to recommend monitoring schemes to ascertain their nature and extent. There are three main reasons for monitoring such impacts:

(i)   To provide an early warning if the actual impact is more severe than expected. It may be possible to implement remedial measures.
(ii) To add to our knowledge of the actual impacts of specific projects on different environments.
(iii) To check the accuracy of predictions and hence improve our predictive abilities for future EIAs.

Selective monitoring of expected adverse impacts, which are not considered significant, may be considered for the same reasons.

2.4.6.8 Impact monitoring/auditing

The monitoring of project impacts is not new. It has been practised for certain major projects for many years. Usually, however, the monitoring of impacts has been implemented as an isolated post-project action having only limited links to EIA work which may have been undertaken previously. The aim of monitoring has been to increase scientific knowledge or ensure that legally enforceable standards are not being exceeded. Also, many impact monitoring strategies, whether or not linked to EIA studies, have been designed badly and the results obtained have been scientifically invalid. They have not been able to indicate that a change in environmental parameters, for example, a fish population in a section of a river, is due to the discharge of effluent from a new installation or to some other factor.

The importance of impact monitoring in relation to EIA has been increasingly recognised. In particular, the activity which has become known as impact auditing or EIA audits (sometime known as post-project analysis) has gained prominence.

Impact auditing is the checking of impact predictions made in EISs to determine their accuracy. This has two specific aspects:

- To determine whether the EIS predicted the complete range of impacts known to have occurred.
- To determine the accuracy of individual predictions contained in an EIS. The basic aim of audits is to improve the ability to predict and understand knowledge of the impacts of different projects in specific environmental settings. Audits are based on the data obtained from impact monitoring both of which activities can assist in project management and improving future EIAs.

EIA is slowly moving away from its focus on the project implementation phase (prior to authorisation) and is now regarded as an activity which has relevance through construction and operational phases. Often EISs are produced at a single point in time and provide only a 'snapshot' of the design of an installation and all EIS predictions relate to this stage in the cycle of the project. In some industrial sectors, for example, the petrochemical industry, technological change is very rapid and designs may alter between production of an EIS and project construction/ operation. To cope with this dynamic situation, EISs should be adaptable and cease being single documents produced solely to assist in project authorisation. They should be revised and updated throughout the life of a project so that the effects of any design changes can be incorporated.

2.4.6.9 Uncertainty

EIA deals with future events and, thus, has to cope with the uncertainties inherent in predictive activity. In the past, uncertainty was ignored, and phrases such as 'will', 'might' etc., were used to qualify in a qualitative manner the uncertainty involved in the predictions. Decision-makers and the public were left to interpret the meaning and significance of such qualitative expressions.

It was realised that there was another 'area' of project assessment, termed 'risk analysis', which dealt more specifically with uncertainty and had developed techniques to estimate the probability of events occurring. Generally, risk analysis is involved in predicting the probability of hazardous events, for example, explosions or the release of toxic gases, and the nature of their consequences. Basically, risk analysis is based on engineering systems and their potential malfunction and then relating the consequences to human health (mortality and morbidity) and structural damage to buildings. Risk analysis does not deal generally with environmental systems, but EIA could be improved by incorporating information on impact probabilities, which could be obtained by the use of risk analysis concepts and techniques (see Chapter 4).

2.5 EIA METHODS AND TECHNIQUES

Having considered the relationship between the various study activities and impact identification, it is important to consider EIA methods which have been developed to help with impact identification. The choice of a suitable method of assessment involves evaluating the needs of the assessor and what each method provides. In general, any method to be good or accurate should be flexible, fairly simple, objective, include all the key environmental issues and be able to identify project generated impacts, and detect sensitive areas. Other factors that need to be considered include the ultimate use of the impacts assessment as a decision or information document. A decision document would require greater emphasis being placed on the identification of key issues, quantification of results, and direct comparison of alternatives. An information document on the other hand, would require a more comprehensive analysis and would focus on interpreting the significance of a broader spectrum of possible impacts.

Table 2.2
Example of simple checklist

1 *Geology*
    1.1 Unique features
    1.2 Mineral resources
    1.3 Slope stability/rockfall
    1.4 Depth to impermeable layers
    1.5 Subsidence
    1.6 Consolidation
    1.7 Weathering/chemical release
    1.8 Tectonic activity/vulcanism
2 *Soils*
    2.1 Slope stability
    2.2 Foundation support
    2.3 Shrink–swell
    2.4 Frost susceptibility
    2.5 Liquefaction
    2.6 Erodability
    2.7 Permeability
3 *Special land features*
    3.1 Sanitary landfill
    3.2 Wetlands
    3.3 Coastal zones/shorelines
    3.4 Mine dumps/spoil areas
    3.5 Prime agricultural land
4 *Water*
    4.1 Hydrological balance
    4.2 Ground water
    4.3 Ground water flow direction
    4.4 Depth to water table
    4.5 Drainage/channel form
    4.6 Sedimentation
    4.7 Impoundment leakage and slope
         failure
    4.8 Flooding
    4.9 Water quality
5 *Biota*
    5.1 Plant and animal species
    5.2 Vegetative community
    5.3 Diversity
    5.4 Productivity
    5.5 Nutrient cycling
6 *Climate and air*
    6.1 Macro-climate hazards
    6.2 Forest and range fires
    6.3 Heat balance
    6.4 Wind alteration
    6.5 Humidity and precipitation
    6.6 Generation and dispersion
         contaminants
    6.7 Shadow effects

7 *Energy*
    7.1 Energy requirements
    7.2 Conservation measures
    7.3 Environmental significance
8 *Services*
    8.1 Education facilities
    8.2 Employment
    8.3 Commercial facilities
    8.4 Health care/social services
    8.5 Liquid waste disposal
    8.6 Solid waste disposal
    8.7 Water supply
    8.8 Storm water drainage
    8.9 Police
    8.10 Fire
    8.11 Recreation
    8.12 Transportation
    8.13 Cultural facilities
9 *Safety*
    9.1 Structures
    9.2 Materials
    9.3 Site hazards
    9.4 Circulation conflicts
    9.5 Road safety and design
    9.6 Ionising radiation
10 *Physiology*
    10.1 Noise
    10.2 Vibration
    10.3 Odour
    10.4 Light
    10.5 Temperature
    10.6 Disease
11 *Sense of community*
    11.1 Community and organisation
    11.2 Homogeneity and diversity
    11.3 Community stability and
          physical characteristics
12 *Psychological well-being*
    12.1 Physical threat
    12.2 Crowding
    12.3 Nuisance
13 *Visual quality*
    13.1 Visual content
    13.2 Area and structure coherence
    13.3 Apparent access
14 *Historic and cultural resources*
    14.1 Historic structures
    14.2 Archaeological sites and
          structures

Table 2.3
Section of a descriptive checklist (from Schaenman, 1976):
Environmental factors

Data required	Information sources/ predictive techniques
Air quality	
Health: change in air pollution concentrations by frequency of occurrence and number of people at risk	Current ambient concentrations, current and expected emissions, dispersion models, population maps
Nuisance: change in occurrence visual (smoke, haze) or olfactory (odour) air quality nuisances, and number of people affected	Baseline citizen survey, expected industrial processes, traffic volumes
Water quality	
Changes in permissible or tolerable water uses and number of people affected – for each relevant body of water	Current and expected effluents, current ambient concentrations, water quality model
Change in noise level and frequency of occurrence, and number of people bothered	Changes in nearby traffic or other noise sources, and in noise barriers; noise propagation model or nomographs relating noise levels to traffic, barriers, etc.; baseline citizen survey of current satisfaction with noise levels

Resource availability including time, skill, money, data, and computer facilities must also be considered. The more quantitative the analysis, the greater the need for resources. Consideration of administrative constraints may also limit choices by the procedural or format requirements of the authorisation mechanisms. These factors need to be evaluated by the proponent of the environmental assessment since he/she is the only one that can best judge the needs and constraints.

Available EIA Methods and Techniques
A large number of EIA methods and techniques have been developed and used in the EIA process. Nearly all the methods are examples or variants of general types which have specific organising principles in common. The main types in common use include checklists, matrices and networks.

Table 2.4
Questionnaire checklist (US Aid Manual)

Terrestrial ecosystems

(a) Are there any terrestrial ecosystems of the types listed below which, by nature of their size, abundance or type, could be classified as significant or unique?

Forest?	Yes	No	Unk
Savanna?	Yes	No	Unk
Grassland?	Yes	No	Unk
Desert?	Yes	No	Unk
Are these ecosystems,			
Pristine?	Yes	No	Unk
Moderately Degraded?	Yes	No	Unk
Severely Degraded?	Yes	No	Unk

(b) Are there present trends towards alteration of these ecosystems through cutting, burning, etc. to produce agricultural, industrial, or urban land?   Yes   No   Unk

(c) Does the local population use these ecosystems to obtain non-domesticated:

Food plants?	Yes	No	Unk
Medical plants?	Yes	No	Unk
Wood products?	Yes	No	Unk
Fibre?	Yes	No	Unk
Fur?	Yes	No	Unk
Food animals?	Yes	No	Unk

(d) Will the project require clearing or alteration of:

Small areas of land in these ecosystems?	Yes	No	Unk
Moderate areas of land in these ecosystems?	Yes	No	Unk
Large areas of land in these ecosystems?	Yes	No	Unk

(e) Does the project rely on any raw materials (wood, fibre) from these ecosystems?   Yes   No   Unk

(f) Will the project decrease use of products from these ecosystems by producing or providing substitute materials?   Yes   No   Unk

(g) Will the project cause increased population growth in the area, bringing about increased stress on these ecosystems?   Yes   No   Unk

Estimated impact on terrestrial ecosystems[a] ND..HA..MA..LA..O..LB..MB..
HB

Disease vectors

(a) Are there known disease problems in the project area transmitted through vector species such as mosquitos, flies, snails, etc?   Yes   No   Unk

Table 2.4 *Continued*

Disease vectors—contd.

(b) Are these vector species associated with:

Aquatic habitats?	Yes	No	Unk
Forest habitats?	Yes	No	Unk
Agricultural lands?	Yes	No	Unk
Degraded habitats?	Yes	No	Unk
Human settlements?	Yes	No	Unk

(c) Will the project:

Increase vector habitat?	Yes	No	Unk
Decrease vector habitat?	Yes	No	Unk
Provide opportunity for vector control?	Yes	No	Unk

(d) Will the project work force be a possible source of introduction of disease vectors not currently found in the project area?	Yes	No	Unk
(e) Will increased access to and commerce with the project area be a possible source of disease vectors not presently occurring in the project area?	Yes	No	Unk
(f) Will the project provide opportunities for vector control through improved standards of living?	Yes	No	Unk

Public health

(a) Are vector-borne diseases an important part of the local public health situation?	Yes	No	Unk
(b) Are there clinics or other disease control programmes in operation or planned for the area?	Yes	No	Unk
(c) Will the project decision result in an increase in disease vector density or distribution?	Yes	No	Unk
(d) Will the project decision result in workers or other persons entering the area with contagious or vector-borne diseases?	Yes	No	Unk
(e) Will the project decision result in clearing operations that could expose workers to disease vectors?	Yes	No	Unk

[a] *Key*: ND, not determinable; HA, high adverse; MA, medium adverse; LA, low adverse; O, low or insignificant; LB, low benefit; MB, medium benefit; HB, high benefit.

2.5.1 Checklists

The term 'checklist' covers a variety of methods having widely varying characteristics and degrees of complexity. However, most share one common feature, that is a list of environmental, social and economic factors which may be affected by a development. The simplest checklist is only able to aid identification of impacts and ensure that impacts are not overlooked (Table 2.2). While such checklists acts as an aide-memoire, additional guidelines are needed to perform other EIA tasks. The checklist was one of the first EIA methods developed and it is still in use in a variety of forms.

An example of a descriptive checklist containing factors to be considered in an EIA is shown in Table 2.3. For each factor, information is provided on data requirements, sources of information and predictive techniques which can be used for assessing impacts. A significant feature of this method is the emphasis placed upon relating all impacts to the people likely to be affected.

Another type of checklist is the 'questionnaire' which presents a series of questions in relation to the impacts of a project. This checklist (Table 2.4) shows two sections (there are 31 in total) from a checklist for assessing rural development projects in developing countries. The questions are listed under generic categories such as 'terrestrial ecosystems' and 'disease vectors'.

Those assessing impacts must attempt to answer the questions in all categories. There are three possible answers. For example, if it was known that an impact was likely or unlikely, then the appropriate answer (Yes or No) would be indicated in the space provided. However, if insufficient evidence was available for a definite response, then the 'Unknown' category would be marked by the assessors. If answers can be obtained, then this checklist provides a scale for designating the estimated impact on disease vectors, for example, an impact is 'high adverse' or 'medium benefit'. The assessment of a project consists of a systematic question and answer procedure with the addition of quantitative and qualitative information on specific impacts when necessary.

Once the first question in a particular category has been dealt with, subsequent questions relate to various aspects of the implications of development for the category concerned. For example, the last question under the category 'terrestrial ecosystems' forces those assessing impacts to consider indirect impacts, that is, the project causing population growth and this in turn affecting the ecosystem under consideration. In addition, the consequences of possible mitigating measures are considered. For example, the last question in the 'disease vectors' category asks whether the project will enable opportunities for control of vectors to be initiated.

The logical progression imposed by such 'questionnaire' checklists and the actions needed to provide the type of information required to answer the question makes this type of approach an advance on other checklist formats.

Checklists of environmental components are limited because they concentrate on only one side of impact identification. Actions associated with a development are not considered explicitly. Impacts cannot be identified in a comprehensive way unless detailed knowledge of the characteristics and actions associated with a project are related systematically to components of the environment. Consequently, checklists of development actions and checklists of environmental components have been brought together in a two-dimensional matrix to aid systematic identification of impacts.

2.5.2 Matrices

The best known of these methods is the matrix developed by Leopold. A section of this matrix which consists of two checklists is shown in Fig. 2.4. A list of development actions is displayed horizontally, while a list of environmental components is displayed vertically. The inclusion of these two checklists in a matrix aids impact identification, because items in one list can be related systematically to all items in the other list to ascertain whether an impact is likely. When all likely interactions between a development activity and environmental components have been identified, the matrix is a useful, easily understood visual summary.

This matrix can be used to measure and interpret impacts as well as identify them, by describing impacts in terms of magnitude and importance on a common 1–10 scale, where 1 is least magnitude or importance and 10 the greatest. The magnitude of an impact is taken to be an expression of its scale, such as the geographical area over which it extends. For instance, a new development may increase ground-level concentrations of SO_2 over a wide area, and this impact receives a magnitude of 8 or 9. This increase may not be thought to cause any harmful effects and will therefore, have a low importance score of 2 or 3.

Assigning scores for magnitude and importance to all identified impacts depends on the subjective views of those assessing a proposal. Detailed guidance is not provided on how these scores can be assigned on a standardised basis, or criteria to assist the assessor decide between particular scores. Since the scores are subjective, they cannot be manipulated arithmetically.

Scores for magnitude and importance should be included in the appropriate cell of the matrix which has been previously identified as representing a likely impact. The cell should be bisected by a diagonal line

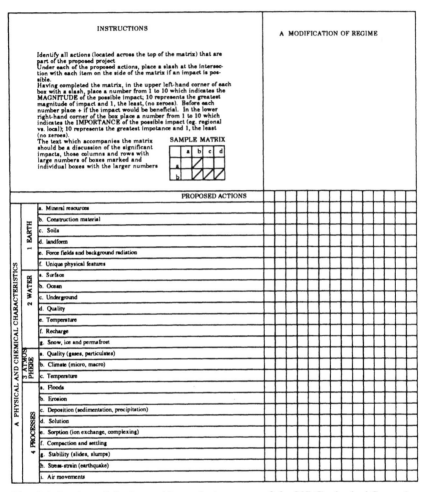

Fig. 2.4. Section of the Leopold matrix (courtesy of the US Geological Survey).

and the score for magnitude placed in the top left-hand corner. The score for importance should be placed in the bottom right-hand corner. These scores can be accompanied by a plus sign ($+$) to indicate whether an impact is beneficial. A completed matrix would consist of a number of cells containing two numerical scores and, in some cases, a plus sign. Although the matrix contains 8800 cells, for most projects, not more than 25–50 cells will be identified as representing impacts. Thus, this matrix can indicate a structured, self-contained format, a considerable amount of information on impacts. However, the subjective nature of the information has to be kept in mind. In addition, this matrix is unable to deal explicitly with the time aspect of impacts.

Despite these drawbacks, Leopold-type matrices have been, arguably, the most commonly used EIA method. Its continued popularity attests to its utility in EIA.

2.5.3 Networks

The first EIA network was developed by Sorensen (1971) to aid planners reconcile conflicting land uses in a section of the Californian coastal zone. Basically, this network consists of a number of linked impacts known to have occurred, from a variety of land uses on the coastal environment, in the past. Figure 2.5 shows a section of the network which deals with water quality.

In the network shown, the chain of impacts consequent upon particular land uses is limited to three links. The first change in the environment is called an initial 'condition' change. For example, land uses such as residential development, commercial services and crop farms will involve drainage improvements which constitute the causal factor resulting in an initial 'condition' change of increased freshwater flow into an estuary. This initial 'condition' change will affect other environmental components, termed 'consequent' condition. In this illustration increased freshwater flow, amongst other effects, can stimulate or increase cliff erosion, and reduce the salinity of the receiving estuary. Although Sorensen included only one 'consequent' condition change in the network it is likely that other condition changes would occur before a final effect was reached.

A network constructed following the principle advocated by Sorensen can be a large and unwieldy visual display. It is possible, however, to computerise a network and the useful potential of computerisation is recognised by Sorensen. Computerisation of a network would enable updating of information on the effects of the land use activities contained in the network. In addition, a computer could select the appropriate section of a network to show expected impacts. This would be an advance as it would overcome the problems of visually distinguishing the appropriate section from a network which existed as a single, large paper chart or a series of smaller charts.

The methods described above are only a few of the varieties available. Nevertheless, they are an important group of methods because, despite a lengthy history in terms of EIA, they are still being revised, amended, developed and used in EIA. One of the main concerns of decision-makers and those carrying out EIA relates to the costs of various EIA methods.

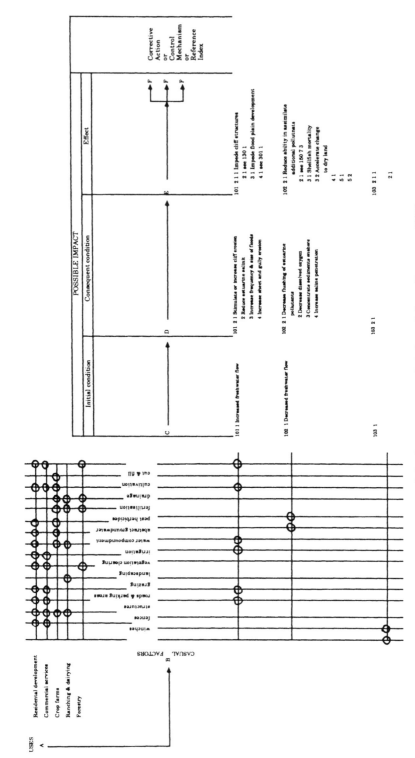

Fig. 2.5. Section of the Sorensen network (adapted from Sorensen, 1971).

Evidence for the relative cost of methods is even more scarce than information on their frequency of use. Most judgements on the cost of using a method are no more than educated guesses based on the requirements of methods. The cost of a method depends, to a great extent, on the number of impacts which have been assessed. This decision is not made by a method, but by those who are involved in identifying and selecting impacts. Most costs result from the prediction and measurement phase of EIA. This is when predictive models, the need for specific base-line data, expert consultants and computer time may be required. All these can be expensive. Therefore, a certain proportion of the costs are determined by factors unrelated to a particular method.

In many cases, methods are not explicit on how a particular impact is to be assessed. Specific techniques are not recommended, nor are other kinds of guidance provided. For example, the questionnaire checklist and the network do not give information on the data needed before a question can be answered or a decision made, whether an impact will result in two/three additional indirect impacts. Decisions on these matters are left to those assessing the impacts. Depending on the decisions made, a particular checklist or network can exhibit a range of minimum and maximum costs. In contrast, there are some methods which have specific data requirements which must be met before the methods can be operated. For example, scaling–weighing checklists need quantitative data on many impacts before they can be used.

There is no universally applicable method which can be applied to all projects in all situations. Further, it is unlikely that such a method will ever be developed because of the variety of situations in which EIA is applied. Consequently, when faced with a particular EIA exercise, a knowledge of available methods with their strengths and weaknesses is useful, as those implementing the EIA will be able to select either the most appropriate method or, as more likely, parts of sections or a variety of methods which can be used together to achieve the particular objectives of the EIA.

At the beginning of this chapter a number of impact characteristics were outlined. The discussion of selected methods has shown that no single method can incorporate all impact characteristics. Networks are good for dealing with direct and indirect impacts. Scaling–weighting checklists are useful for comparing all impacts from a number of alternatives. Overlays have the ability to deal with the spatial aspects of impacts and with linear projects. The problem to be faced with all methods is the complexity that would result if all these impact aspects were to be incorporated in a single method.

Matrices, overlays and networks are among the most popular methods. Their limitations are well known, but their advantages are considered to outweigh any disadvantages. In fact there is a growing realisation that a combination of simple interaction matrices and networks can be a very effective means of identifying impacts. In addition, they can guide the assessment process and present information on impact in an easily understandable format. A simple interaction matrix can be used to identify likely first-order, direct impacts. Once this has been done a network approach would allow the range of indirect impacts to be identified. Consideration of the various characteristics of impacts such as reversibility and probability would provide a useful input into the formulation of a network and the written description of the impacts assessed.

None of these methods possesses the main advantage of the scaling–weighting checklist. This method fulfils a need often ignored by other methods, namely the desire of many decision-makers to be faced with an easy decision, especially when comparing a complex variety of impacts from a number of alternatives. Also, should decision-makers be sufficiently interested to test different assumptions, then such methods can quickly show the outcomes.

Two other factors may account for the existing and likely future popularity of these methods. First, many countries have political systems which exhibit or assume general consensus. It is likely that these methods will be popular in such decision-making contexts. Also in many countries, EIAs are undertaken by engineers and technologically trained people who have an affinity for the use of quantitative aids in their work.

For these reasons it is likely that index methods will have an assured future, despite their major failings in the way they treat the environment as a series of discrete, separate components.

The debate on the utility of various methods tends to be hypothetical and philosophical because no information exists on the actual operational performance of many types of methods. Such information could be obtained by assessing a selection of projects by different methods or approaches and to compare not only cost/resources requirements, but also the outcomes to see if choice of method has any effect on the selection of the environmentally preferred alternative. Such a study would be of great benefit for those likely to be undertaking EIAs in the future.

2.6 EIA AND POLLUTION CONTROL

The EIA process will involve the collection of information relating to the release of pollutants to air and water, including the scale, amount and type of effluents. In addition, it is necessary to have data on emission characteristics, for example, temperature, exit velocity and concentration. With this type of information and knowledge of prevailing environmental conditions, such as ambient noise levels and habitat types, it is possible to make predictions of likely impacts. Once impacts have been predicted, they can be considered for their relative importance. Some will be thought to be more significant than others. Consequently, the results of an EIA will pinpoint those impacts which are likely to have most adverse or beneficial effects. In terms of environmental protection only adverse environmental impacts will be important. From knowledge of the most significant impacts, and data obtained on the nature of emissions and existing environmental characteristics, it is possible to use EIA as a means of controlling the emission of pollutants. Control cannot be implemented directly but information gained from EIAs can be used to guide discussions of design changes aimed at pollution control. EIA can also plan an indirect role in pollution control through the identification and establishment of monitoring schemes. Potentially harmful impacts identified by EIA can be monitored. Data obtained from monitoring can help ensure that adverse effects are detected at an early stage so that remedial action can be taken before changes become a serious problem requiring time-consuming and, in many cases, expensive mitigating action. The utility of predictive techniques used in an impact assessment can only be improved if results obtained during an assessment are checked with information obtained once the development becomes operational.

REFERENCES AND BIBLIOGRAPHY

Advanced Studies in Science, Technology and Public Policy (1982) *A Study of Ways to Improve the Scientific Content and Methodology of Environmental Impact Analysis*, Final Report to the National Science Foundation on Grant PRA – 79–10014. Advanced Studies in Science, Technology and Public Policy, School of Public and Environmental Affairs, Indiana University, Bloomington, Indiana.

Aird, R. (1982) *Environmental Monitoring of Project Impacts; Regulatory Requirements and Developing Concepts and Approaches*, Graduate Student Monograph 15, Calgary, Petro Canada.

Barnthouse, L.W. et al. (1984) Population biology in the courtroom: the Hudson River controversy, *Bio Science*, **34**, 14–19.

Battelle Institute (1978) *The Selection of Projects for EIA*, Commission of the European Community Environment and Consumer Protection Service, Brussels.

Beanlands, G.E. and P.N. Duinker (1983) *An Ecological Framework for Environmental Impact Assessment in Canada*, Federal Environmental Assessment Review Office, Ottawa.

Bishop, A.B. (1975) Public participation in environmental impact assessment. In *Environmental Impact Assessment* (M. Blisset, ed), Lyndon B. Johnson School of Public Affairs, University of Texas at Austin, pp. 219–236.

Canter, L. (1983) Methods for Environmental Impact Assessment: Theory and Application (Emphasis on Weighting–Scaling Checklists and Networks). In *Environmental Impact Assessment* (PADC Environmental Impact Assessment and Planning Unit, ed), Martinus Nijhoff, The Hague, pp. 165–234.

Dee, N. et al. (1971) Environmental evaluation system for water resource planning, *Water Resource Res.*, 9, 523–35.

Dooley, J.E. and R.W. Newkirk (1976) *Corridor Selection Method to Minimise the Impact of an Electrical Transmission Line*, James F MacLaren Ltd., Toronto.

Duke, K.M. et al. (1977) *Environmental Quality Assessment in Multiobjective Planning*. Battelle Columbus Laboratories, Columbus, OH.

Duke, K.M. et al. (1979) *Environmental Quality Evaluation Procedure Implementing Principles and Standards for Planning Water Resources Programs*, Draft Report, US Water Resources Council, Washington, DC.

Eagle, P.F.J. (1981) Environmentally Sensitive Area Planning in Ontario, Canada, *APA Journal* July 1981, 313–323.

EAP (1976) *Guidelines for Preparing Initial Environmental Evaluations*, Environmental Assessment Panel, Ottawa.

FEARO (1978) *Guide for Environmental Screening*, Federal Activities Branch, Environmental Protection Service and Federal Environmental Assessment Review Office, Ottawa.

FEARO (1980) *Environmental Assessment Panels — What they are — what they do*, Federal Environmental Assessment Review Office, Ottawa.

Fabos, J.G.Y. et al. (1978) *The METLAND Landscape Planning Process: Composite Landscape Assessment, Alternative Plan Formulation and Plan Evaluation: Part 3 of the Metropolitan Landscape Planning Model*. Research Bulletin No 653, University of Massachusetts Agricultural Experiment Station, Amherst, MA.

Galloway, C.E. (1978) *Assessing Man's Impact on Wetlands*, Report No 78–136, University of North Carolina Water Resources Research Institute, Raleigh, NC.

Green, R.H. (1979) *Sampling Design and Statistical Methods for Environmental Biologists*, Wiley, Chichester, UK.

Grima, A.P. (1977) The Role of Public Participation in the Environmental Impact Process. In *Environmental Impact Assessment in Canada Processes and Approaches* (M. Plewes and J.B.R. Whitney eds), Environmental Studies, University of Toronto, Toronto, pp. 61–76.

Haimes, Y.Y. et al. (1975) *Multiobjective Optimisation in Water Resource Systems: The Surrogate Worth Trade-Off Method*, Elsevier, New York.

Heer, J.E. and D.J. Hagerty (1977) *Environmental Assessment and Statements*, Van Nostrand Reinhold, New York.

Hill, M.A. (1968) A goals-achievement matrix for evaluating alternative plans, *J. Am. Inst. Planners* **34**, 19–28.

Holling, C.A. (ed) (1978) *Adaptive Environmental Assessment and Management*, Wiley, Chichester, UK

Leopold, L. et al. (1973) *A Procedure for Evaluating Environmental Impact*, US Geological Survey Circular 645, US Geological Survey, Washington, DC.

Leonard and Partners (1977) *Belvoir Prospect, Surface Works Report Volume 2 Plans*, Leonard and Partners, Croydon.

McHarg, I. (1968) *A Comprehensive Highway Route-Selection Method*, Highway Research Record No 246, Highway Research Board, Washington, DC.

Manning, W. (1913) The Billercia town plan, *Landscape Architect.*, **3**, 108–18.

Munn, R.E. (ed) (1979) *Environmental Impact Assessment: Principles and Procedures*, SCOPE Report 5, 2nd edition, Wiley, Chichester, UK.

Organisation for Economic Cooperation and Development (1979) *Environmental Impact Assessment*, OECD, Paris.

O'Riordan, T. (1980) *Environmental Impact Assessment and Policy Review*, Paper presented at the IDG Conference, Lancaster University, January 1980.

PADC (1981) *Manual for the Assessment of Major Development Proposals*, HMSO, London, UK.

Sassaman, R.W. (1981) Threshold of Concern: A Technique for Evaluating Environmental Impacts and Amenity Values. *J. Forestry*, **79**(2), 84–6.

Schaenman, P.S. (1976) *Using an Impact Measurement System to Evaluate Land Development*, The Urban Institute, Washington, DC.

Sorensen, J.C. (1971) *A Framework for the Identification and Control of Resource Degradation and Conflict in the Multiple Use of the Coastal Zone*, unpublished Master's Thesis, Department of Landscape Architecture, University of California at Berkeley, Berkeley, CA.

Suter, G.W. et al. (1986) Treatment of risk in environmental impact assessment. In *Risk and Policy Analysis Under Conditions of Uncertainty*, CP–86–26 (C.T.Miller, P.R. Kleindofer and R.E. Mann eds), International Institute for Applied Systems Analysis, Laxenburg, Austria.

Solomon, R.C. et al. (1977) *Water Resources Assessment (WRAM)* — *Impact Assessment and Alternative Evaluation*, Technical Report No Y–77-1, US Corps of Engineers, Vicksburg, MS.

Sondheim, M.W. (1978) A comprehensive methodology for assessing environmental impacts, *J. Environ. Manage.*, **6**(1), 27–42.

Thor, E.C. et al. (1978) Forest environmental impact analysis — a new approach, *J. Forestry*, November, 723–5.

Tomlinson, P. and Atkinson, S.F. (1987) Environmental audits, proposed terminology, *Environ. Monitor. Assess.*, **8**, 187–98.

United Nations Environment Programme (1978) *Draft Guidelines for Assessing Industrial Environmental Impact and Environmental Criteria for the Siting of Industry*, UNEP Industry and Environment Office, Paris.

United Nations Environment Programme (1980) *Guidelines for Assessing Industrial Environmental Impact and Environmental Criteria for the Siting of Industry*, UNEP Industry and Environmental Guideline Series, Vol. 1, UNEP Industry and Environment Office, Paris.

US Agency for International Development (1980) *Environmental Design Consideration for Rural Development Projects*, US Agency for International Development, Washington, DC.

US Department for Housing and Urban Development (1975) *Interim Guide for Environmental Assessment*, US Department of Housing and Urban Development, Washington, DC.

US Fish and Wildlife Service (1980) *NEPA Planning and Documentation FWS Handbook*, Office of Environmental Coordination, US Fish and Wildlife Service, Department of Interior, Washington, DC.

Wood, W.M. (1978) Public involvement techniques utilised in highway transportation planning. In *Environmental Assessment: Approaching Maturity* (S. Bendix and H.R. Graham eds), Ann Arbor Science, Ann Arbor, MI, pp. 205–13.

Chapter 3

Toxicology, Ecotoxicology, Environmental Epidemiology and Human Health

3.1 INTRODUCTION

The thalidomide disaster in the early 1960s stimulated the development of modern toxicology. Until that time toxicology was primarily concerned with medicinal products, poisons and poisonings. It has now come to signify the science of poisons, or the study of the adverse effects of chemicals on living tissues.

The tremendous growth of the chemical industry from a pre-war production of about one million tons, to the hundreds of millions of tons in recent years has meant that today, there is an extensive class of substances that had not existed in the environment until their creation by human scientific and industrial effort. These are the synthetic chemicals that play an increasingly dominant role in human affairs throughout the entire world. Their number, importance and pervasiveness are still not fully appreciated.

The phenomenal increase in the use of pesticides and the ubiquitous use of plastics in every day items were based on the development of thousands of new chemicals by research laboratories in the chemical industry. It has been estimated that there are now some 100 000 chemical entities of which about 40 000 are in common use.

The rapid development of new chemicals to meet the needs in agriculture, industry and the household were accompanied by some disbenefits for human, animal and environmental health. It was the dramatic emergence of some of these hazards and the urgent calls for action by public health authorities and other national and international bodies with responsibilities for health that more or less forced the evolution of modern toxicology. Its development now has two broad strands of effort, one being the increased public awareness of the possible dangers of chemicals and the subsequent efforts by responsible authorities to devise control measures, the second being the enormous progress made in the understanding of the molecular mechanisms of cellular functions, particularly those concerned with genetic mechanisms.

At the public health level there were clear indications of a need for some legislative measures to ensure safety in the use of dangerous chemicals from the very earliest developments in the modern chemical industry, namely the dye and coal tar industries of the early part of the century. Unfortunately, it took a series of catastrophic events during the past few decades to bring about a recognition that some effective control measures were needed.

3.2 TOXICOLOGY AS A MULTIDISCIPLINARY SCIENCE

3.2.1 The Contribution of the Basic Sciences

Although toxicology must still be regarded as an evolving subject, the results of toxicological studies have become of importance to professional bodies for widely varying reasons. It must be emphasised that the simple question — how toxic is this chemical? — can rarely have a simple answer. An answer will depend not only on the nature of the chemical but also, among other considerations, on the interests of the person asking the question, the circumstances of the exposure, the kind of populations exposed as well as matters such as age, sex, species and genetics. For any given set of variables the assessment of the probability of risk will often be quite different.

Anatomically, the body can be viewed as contacting the environment, or aspects of it, via three principal surfaces: the skin, the respiratory and the gastrointestinal systems. Thus for any substance to enter the tissues of the body it must cross one or more of three barriers. The commonest way for a toxic substance to enter the body is by ingestion, usually as a contaminant of food, or by breathing.

The EIA has, in toxicology science, a valid qualitative and quantitative instrument to evaluate risks for people living in a possible affected environment. For this purpose, many standardised toxicological procedures are available and a correct application of them may help managers to take decisions to protect the population from health hazards or to regulate pollution in order to reach tolerable levels that result in an 'acceptable' degree of risk for the population. Further developments in this discipline (e.g. DNA adduct analysis, genetic epidemiology, etc.) will contribute to improve knowledge on possible effects for human health, thus increasing the importance of toxicology and its contribution to EHIA.

3.2.2 Practical Considerations
3.2.2.1 The problems
By the 1970s there was ample verification that new chemicals could be harmful and much had been learned about the kinds of damage that could result from exposure. The discoveries and revelations of the past decade in particular brought about a fundamental shift in toxicological thought and hence in subsequent toxicological developments, both in the public health sphere and in the laboratory. What was clear was that health problems would arise from long-term low level exposures. The classic example of unravelling the dangers of cigarette smoking and particularly the different kinds of diseases that could be promoted, has had a profound influence on attitudes to the reality of the dangers of long-term exposures and the need for preventive measures. Cancer was an obvious long-term low level exposure effect, once it was realised that chemicals could be carcinogens. This resulted in the rapid development of mutagenicity tests in the hope that they could be used as a screen for filtering out potentially dangerous molecules before they reached the market place. Carcinogens, because of the great public concern, received much attention, however evidence was also available on a number of other chronic diseases in which chemical exposures were implicated.

The study of teratogens showed that there were other possibilities of interference with reproductive processes and thus there was a need to broaden the considerations to the whole reproductive cycle. Sterility for example is one possible consequence, particularly in relation to agricultural chemicals. Other specialised areas of toxicology, such as neurotoxicity, behavioural toxicology and immunotoxicology were recognised and added to the classic areas of hepatotoxicology, nephrotoxicology and toxicokinetics. The health issues that have emerged as a result of the explosion of knowledge about the hazards of chemicals have become matters of priority among those organisations, both national and international, that have responsibility for public health. Unfortunately, the tremendous growth of the chemical industries during the past decades has not been paralleled with the development of the appropriate toxicological expertise, either in manpower or technology. Thus the situation has developed where the awareness of the problems is growing but the resources to deal with the problems lag far behind. This obviously must have serious practical consequences.

The problem facing those responsible for public health matters is simply that chemicals could pose serious health hazards, but with tens of thousands of chemicals in common use, and with totally inadequate resources, what could be done? There is a well-established system for the testing of drugs for unwanted toxic effects, but the problem with chemicals is somewhat different. Drugs are usually prescribed for specific reasons and for limited durations. They are regarded as dangerous unless used as instructed and there are presumed benefits to the patient. The exposure possibilities of individuals to chemicals are varied and often a person is unaware of an exposure. At one extreme, there is the situation in the workplace where the use of a dangerous chemical may be unavoidable. However, it is possible to work with a chemical of known toxic properties, by containment at every stage of the manufacturing process. At another extreme, the safety assessment of a food additive, to which populations may be exposed for a lifetime, quite obviously, must be particularly stringent. Thus the problems of chemical safety are those of risk assessment and risk management, and risk assessment depends on assembling data about toxic properties. This is the responsibility of the toxicologist.

3.2.2.2 A strategy for toxicity testing

The Organisation for Economic Cooperation and Development (OECD) and the European Economic Community (EEC) and a large number of national agencies have recommended that the testing for the toxicity of chemicals should be conducted in a stepwise fashion with the addition of more stringent test requirements as the exposure possibilities increase. The operation of this in practice can be seen from the way in which the Commission of the European Community have implemented their test requirements. In brief the requirements for a new chemical are specified as a dossier of information, an essential part of which is a technical dossier containing the results of specified physicochemical, toxicological and ecotoxicological tests that constitute the so-called base set of data. As the tonnage released to the market increases, so there will be an increasing requirement for test data.

The base set requirements for toxicity data and the reasons for their selection are important matters to consider. The first step and a most important one, is a single dose (or acute) toxicity study. For practical reasons this is carried out, as are most tests in the base set, in small rodents. This establishes any signs of toxicity and their time course, the organs affected (target organs) and some idea of potency as a toxic agent. This is expressed in terms of an estimate of the lethal dose (LD_{50}). This is a basic fact about any chemical. International regulations governing the classification of the toxicity of a chemical so that appropriate labelling and handling conditions can be set are based on the LD_{50} value. A precise value of the LD_{50} is not a matter of concern, what matters is whether the value is in micrograms, milligrams or grams per kilogram. Strychnine, for example, is an extremely toxic substance and its LD_{50} is 2 mg/kg, botulinum toxin has an LD_{50} of a hundredth of a microgram and ethyl alcohol 10 g/kg. The LD_{50} will vary according to the route of entry and so this is selected on the basis of the known or intended use. The acute toxicity studies also provide the basis on which the consequences of accidental exposure can be assessed and they also determine the dose level at which toxic effects can be expected, a key element in the planning of the subsequent longer term studies.

Chemicals must be handled from the earliest stages of their development and thus it is important to know from the outset if exposure can cause local damage to the eye or the skin and thus eye and skin irritation studies form part of the base set of acute toxicity studies.

The implementation of the control measures along the lines recommended by the OECD has been a difficult task for national authorities. It has meant that toxicology in recent years has been dominated by questions about the testing of chemicals. Unfortunately, lawyers and legislators expect more of biological test systems than can be delivered. It is all too easy for those involved at the legislative level of toxic chemicals control.

Thus it is important to understand just what a 'test' is in a toxicological evaluation process. The classical tests in the base set, namely the acute, subacute and chronic tests are most complex procedures that can only be conducted in highly specialised laboratories.

3.2.2.3 The development of new test procedures

It is useful to consider the steps in the evolution of a test that has reached the status of incorporation into some legislative requirements. The process starts with a research scientist, who devises a procedure that shows promise for the study of some biological property. In pharmacology this would be known as a preparation. Many of these preparations progressed to well-established biological assay procedures for molecules that could interfere with functions, such as the curare-like molecules that can block the action of transmitter agents at neuro-muscular junctions. In selecting test procedures that might be included in a collaborative study, it is useful to consider the process of exploitation as a series of steps which are set out as follows:

- A research paper that describes a procedure that could be useful for the detection of a particular property with a few illustrative chemicals.
- The idea is taken up by other laboratories who confirm the usefulness of the procedure and add more chemicals to the data base.
- The procedure, possibly with modifications, begins to be discussed as a feasible new test at national scientific meetings and a small inter-laboratory trial is arranged by interested research scientists.
- The general success of the three previous steps generates some international interest and more scientific papers begin to appear and the data base begins to assume a significant size. At this stage the procedure is accepted among the scientific community as a novel test.
- The novel test procedure begins to enter the 'test dimension' in toxicological and legislative circles. Good laboratory practice and quality assurance become important. Questions of protocol, sensitivity and specificity begin to be examined and a collaborative study is initiated with some government agency support. At this stage much of the work needed to answer these questions will be undertaken by commercial toxicological test laboratories.
- The novel test procedure has by now been widely discussed and a number of controversial issues will have emerged, particularly concerning discrepant results and the possibilities of false negative and false positive results, together with debates about the relevance of the findings to human disease.
- This is the stage when the results from the test procedure are beginning to be considered informally in the assessment of the health risks of chemicals by legislative authorities. Problems of protocol 'robustness' of the test, sensitivity and specificity become important and international collaborative efforts to answer these and other questions that have arisen in the course of the development of the test are set in train.
- The culmination of all these activities is an international consensus that the test is not acceptable, or that it can be accepted as a special test to be deployed in the final stages of hazard assessment in suitable circumstances or, the ultimate accolade, adoption as a test for the general screening of a chemical for a specific hazardous property.
- The test becomes incorporated by government authorities in the chemical dossier requirements for the registration of a new chemical.

For over two decades there have been many proposals for new test procedures. In the short-term test field, about one hundred can be identified, but only ten are at present widely used.

3.3 APPLICATION OF TOXICOLOGY TO ENVIRONMENTAL PROBLEMS

The evolution of toxicology has been discussed primarily with relation to human affairs. The basic toxicological knowledge generated because of the concerns about human health have relevance to all living matter, however 'other living matter' comes well down the priority list for action. Mention has been made about the discovery of the world-wide contamination with DDT and the PCBs and how this initiated an awareness of the possibilities of chemicals having harmful effects on the environment. The follow up of these concerns has developed into the subjects known as *environmental toxicology* and *ecotoxicology*. These subjects encompass effects whose ultimate health impact is potentially every bit as serious as the direct effects of chemicals on the human species. But, as yet, few humans apprehend how closely they are, as a whole, linked with and thus dependent upon the balanced operation of all the living systems on this little planet.

The basic concerns of environmental toxicology can be viewed as being somewhat similar to those of conventional toxicology. Metabolic processes become chemical changes brought about by factors such as ultraviolet light. These reactions can produce a product that has a greater toxicity than the original. Parallels can be drawn about the elimination processes in that chemicals have to be degraded (that is 'metabolised') to harmless products. Target organs in animals become the target organisms in the environment. Here, however, there is as yet no adequate pathology available. Model systems have to be developed in which the effects of chemicals can be studied on a limited number of interacting species. Species have to be identified that are particularly susceptible to classes of chemicals, so that they can be used as indicator organisms. Some progress has been made in this area in that some plant species have been used to detect the presence of mutagenic chemicals in the environment near factories. Sample species have to be selected for classical toxicity studies so that safe limits (no effect levels) can be set for chemicals that must be discharged to the environment, particularly into waterways. Such discharges and their fate is perhaps the circulatory equivalent in animal studies. The problems of acid rain illustrate how difficult and controversial this can be.

It can be seen that the science of toxicology has not developed in quite the way that other sciences have grown. It is rather similar to medicine in that there are academic and practical aspects with profound social implications. Thus, it is particularly important to see the subject and its developments in perspective so that no one aspect comes to dominate to the detriment of the whole. The worry about toxicology is that the pressure on authorities to take action may bring about the adoption of test procedures before they have been fully validated. It takes a long time to get a test into a legal set of requirements and it takes even longer to get it out. If the test proves to be ineffective then there could be a considerable wastage of resources and diversion of effort away from real issues of concern. Resources will always be limited and much thought must go into the political decisions about the allocation of funds and the support for appropriate legislation. To this end it is important that those scientists involved 'at the sharp end' in these matters should endeavour to develop better powers of communication with the politicians and their civil servants. It is they after all, who are ultimately responsible for effective action in matters that bear on the public interest.

3.3.1 Qualitative and Quantitative Toxicological Risk Assessment
The discipline of *toxicology* can contribute information and procedures to assess human risk in an EHIA process. In the multi-step scheme for the 'health oriented' shown below, both qualitative and quantitative toxicological risk assessment play a key role.

(1) Identification of health hazard sources
(2) Quantitative assessment of pollution loads and emission factors
(3) Identification and assessment of serious accident risks
(4) Health oriented study of the potentially affected systems
(5) Estimation of environmental pollution levels and definition of possible exposure scenarios
(6) Assessment of adverse effects connected with the identified exposure scenarios

Risk assessment for human health has been defined as a scientific process, to be clearly distinguished from the risk management process. The scientific process consists of three objective components as follows:

- *Hazard identification*: which is the qualitative assessment phase of the process. It is based on the evaluation of all available data to characterise health risks for the exposed population.
- *Exposure assessment*: which is aimed at evaluating the type, magnitude, time and duration exposure and, to the characteristics of an exposed population in terms of the number of people exposed, sensitive groups and so on.
- *Dose-response assessment*: which is the quantitative part of the whole process. It estimates the relationship between the dose of a substance and the incidence of the adverse health effect.

Each of these components is a basic and necessary step in the process of assessment of risks associated with chemicals. The qualitative and quantitative aspects of toxicological risk assessment in EHIA are analysed in Chapter 4.

3.4 PRINCIPLES AND TECHNIQUES OF EPIDEMIOLOGY IN ENVIRONMENTAL HEALTH

3.4.1 Epidemiological Definitions

Epidemiology may be defined as the study of diseases or other health-related events in communities. It thus differs from clinical medicine where individual patients provide the target for attention. Being population-oriented, number-counting is a central component. The numbers can be births, deaths, cases of illness, cases of recovery and so on. The numbers of new cases, for instance of deaths, during a specified period in a given population constitute the *incidence rate*; the numbers present at a given point in time, or during a specified period, constitute the *point prevalence* or *period prevalence*, respectively.

Epidemiology differs also from clinical medicine in having preventive medicine as its major objective, rather than curative or palliative medicine. Nevertheless, clinical and epidemiological practitioners have many concepts in common, amongst which is the importance placed on obtaining an accurate history of the case, whether this be for a sick person or a sick community. In both types of investigation, the questions about the disease which can bring a diagnosis are the simple ones: What? Who? Where? When? and Why?

It is the variations in disease rates between populations subject to different epidemiological variables — personal characteristics (age, sex or social and ethnic backgrounds, for instance), at different places (whether nations or villages) and at different times (whether days or decades), which provide the clues for diseases sought in preventive medicine environmental health.

3.4.2 Epidemiological Variables

3.4.2.1 Personal characteristics

Age: that disease and death are more frequent at the extremes of the human lifespan is old news. But perhaps less well known are the diseases which attack ages not normally associated with illness. Hodgkin's disease and breast cancer, for example, with their two-pronged assaults, striking young adults as well as people at late middle age. A *biphasic* incidence by age raises the possibility that the disease may have different causes in the two age groups. Again, when a disease affects an age group of people where it is not usually found, unusual environmental causes — perhaps in the occupational environment — may be operating.

Sex: some diseases lay unequal claims on males and females. Males, for example, are more prone to respiratory cancer, ischaemic heart disease, inguinal hernia and peptic ulcer. Females more frequently develop thyrotoxicosis, diabetes, gall bladder diseases and arthritis of the hip and pelvis. Exceptions to these general rules, such as the almost equal incidences for respiratory cancer in south-east Asia, strongly suggest that different types of cause may exist in the different ecological environments.

Social and ethnic backgrounds: most frequently, the higher the social class the lower the mortality from disease. This phenomenon of the social class gradient appears with the common diseases of the circulatory, respiratory and gastrointestinal systems; other diseases such as lymphomas and leukaemias show no gradient, whereas some rarer diseases (cancers of the brain, eye and testis, for example) may show an inverse gradient. Studies of ethnic groups within a country have demonstrated many differences in the disease rates.

Migration studies have shown that rates of disease in the incoming ethnic groups begin to approach the rates of the host population with succeeding generations; gastric cancer amongst Japanese migrants, for instance. Migration studies showing these changes in rates constitute strong arguments for the predominant role of environmental factors over genetic factors in the causation of many diseases.

3.4.2.2 Geographical differences
On a global scale, the registration rates for gastric cancer show a gradient declining from east to west in the northern hemisphere, whereas the gradients for colon and breast cancers are in the contrary direction. Again on the international scale, respiratory cancer in Scotland has been consistently the highest in the world for many years, with England and Wales close behind; yet New Zealand, although largely peopled by immigrants from the United Kingdom, shows a mortality which is far lower.

Within a country, the geographical distribution can vary appreciably. In a recently published Atlas of Mortality for Scotland, cardiovascular diseases showed a consistent predilection for the south-west of the country, whereas the north-east was far healthier for this category of disease. The reason for that pattern is not known. Within the industrialised central belt of Scotland, on the other hand, the numbers of communities with high mortalities from bronchitis and respiratory cancer bore witness to the pathogenic actions of industrial air pollution on the lungs of the populations living there. In China, the Middle East and Africa, exceptionally high rates of oesophageal cancer can be found in districts adjoining other districts where the rates are normal or low.

At the level of a neighbourhood, high mortality and morbidity in a community can sometimes be related to environmental pollution from a local course.

The cases of community poisoning by dioxin-related chemicals at Seveso, and by toxic gases at Bhopal, provided acute examples of industry-related epidemics. The tragedies of Minamata and Itai-Itai diseases in Japan, caused respectively by mercury and cadmium, were more insidious.

Whatever the geographical scale, the presence of a disease, either where it is not expected at all or where its rate differs appreciably from the rates in neighbouring localities, is of central relevance to environmental epidemiologists searching for causal factors for diseases.

3.4.2.3 Time trends

Changes in disease rates attract attention, particularly when the disease has a rapid onset and a colourful public image. Notorious examples were provided by the historic epidemics of the infectious diseases such as the Black Death which swept westwards and northwards across Europe, and probably triggered by the successful early use of germ warfare, when besieging Tartars lobbed corpses of plague victims by catapult into an embattled Crimean port; the Asiatic cholera, which moved more slowly but equally inexorably from its homeland in the areas of Calcutta along routes of trade and pilgrimage into Europe. Dramatic epidemics such as these have naturally galvanised investigations and successful preventive measures.

Slower-moving changes acting over years or decades, however, have also generated investigations: the appearances of Minimata and Itai-Itai diseases; the study of phosphorus necrosis of the jaw ('phossy jaw') in workers at Bryant and May's match factory in London around the turn of the century, caused by their use of yellow phosphorus; the rise in respiratory cancer rates in the United Kingdom and other European countries during the present century, now known to be largely due to cigarette smoking; the declining rates (as yet not convincingly explained) over recent decades for gastric cancer in many western countries, and for anencephaly in Holland.

3.4.2.4 The 'causal' variable

In real life, variables do not exert their actions on health in isolation. For example, people residing in some geographical areas within a country or a city may differ from those in other areas by occupational group or perhaps by ethnic background. These differences could be further bound up with differences in diet or in smoking habits. Thus, the complexities of the interacting networks of variables challenge the epidemiologist to identify a single causal variable.

The epidemiologist must constantly make allowances for those common pitfalls which can produce apparent differences in disease prevalence: differences in, for example, the accessibility and utilisation of care, and in the precision of reporting or diagnosing diseases.

When unexpectedly high rates of disease have been found, and a suspicious environmental factor has been identified, can it be certain that the two phenomena were linked causally? The short answer is no. However, there is a list of criteria which, if at least partly satisfied, suggest that the link might be causal:

- *strength of association*: how statistically significant are the abnormalities?
- *plausibility*: is there a credible scientific argument?
- *biological gradient*: does a greater exposure to the environmental factor increase the health abnormality?
- *temporality*: did the environmental factor precede the health abnormality?
- *specificity*: are the environmental factors and health effects identifiable as clearly isolated?
- *consistency*: can the epidemiological association be reproduced elsewhere?
- *experimental evidence*: can the toxicological link be demonstrated under laboratory conditions?

These criteria (although none singly provide unequivocal evidence of causation) are among the best guidelines available in the search for environmental causes of disease.

3.5 ENVIRONMENTAL HEALTH: OCCUPATIONAL HYGIENE

Occupational hygiene is the applied science concerned with the identification, measurement, assessment of risk and control of environmental

Table 3.1
Identification/classification of occupational hazards

Chemical	Physical (energy)	Biological
Liquid	Thermal	Bacteria
Vapour	Noise and vibration	Yeasts
Gas	Barometric	Fungi
Dust	Radiation	Spores
Fume	UV	Insects
Mist	Visible	
	IR	
	Laser	
	Microwave	
	Mechanical	
	Electrical	

factors arising in the workplace which may affect the health, comfort or efficiency of workers or members of the community.

The role of the occupational hygienist includes the following.

3.5.1 Identification of Harmful, Unpleasant or Uncomfortable Factors

The chemical, physical and biological factors of environmental concern are summarised in Table 3.1.

Identification of potential problems involves:

- the study of existing and proposed plant, equipment, materials used, products and by-products, production processes and general working conditions with review of appropriate engineering design documents;
- the study of the relationship of the workplace to the neighbouring community, including location of buildings, topography and meteorological data;
- the study of products and their intended methods of use, to identify potential hazards and safe working procedures for customers;
- having contact and discussion with management and employees at all levels concerning the acceptability of the working environment;
- the collection and study of up-to-date information in the field of occupational health and hygiene.

Identification of these problems are best accomplished by means of a walk-through occupational hygiene survey.

3.5.2 Measurement of Relevant Environmental Factors

(1) Environmental monitoring may be required for several reasons:
   - (i) to identify environmental contaminants or their sources;
   - (ii) to assess personal exposures and/or background levels;
   - (iii) to assess the efficiency of control measures;
   - (iv) for epidemiological research purposes.

(2) The sampling strategy will depend upon the purpose of the investigation and the variability of the environmental factor with time, location, job, etc. The number of measurements required and their distribution between persons, workplaces and occupations need to be assessed statistically in relation to the precision required.

(3) The selection of instruments and methods must take account of the potential hazard. For example, a short-term measurement or continuous direct readings would be appropriate where an acute hazard is suspected (e.g. gases such as formaldehyde or ozone where a maximum or ceiling value applies). Alternatively, a long-term or shift average measure is relevant in the case of silica or other dusts concerned in pneumoconiosis related diseases.

(4) Personal samples taken on the worker are normally preferred for assessing personal exposures. Static samplers placed in the workplace environment may be relevant when measuring background conditions or the efficiency of control equipment.

(5) Sampling procedures must also take account of analytical requirements and the precision required.

(6) Analysis of the materials used in the work processes may be required.

(7) It may also be desirable to monitor biological samples of blood and urine from exposed persons.

3.5.3 Assessment Risk

(1) Assess the validity of the results.

(2) Compare with recommended standards in respect of short-term and long-term exposure (e.g. HSE Occupational Exposure limits, guidance notes, codes of practice; ACGIH threshold limit values, short-term exposure limits, ceiling values).

3.5.4 Control of the Environment

(1) Recommend methods of environmental control to protect workers and the general public.
   — substitution
   — segregation

— complete enclosure
— partial enclosure
— exhaust ventilation
— personal protection

(2) Assist in the preparation of health and safety data sheets, warning notices and other information to warn employees and the public of any necessary precautions for the safe use of materials products.

(3) Participate in the preparation of regulations, standards and safe working procedures.

3.5.5 Research and Development

(1) Participate in research projects to identify and control new hazards. Identify development needs for hazard evaluation and control technology.

(2) Participate in epidemiological studies of occupational diseases in relation to environmental factors.

REFERENCES AND BIBLIOGRAPHY

Abramson, J.H. (1984) *Survey Methods in Community Medicine*, Churchill Livingstone, Edinburgh.

DHHS (1985) *Risk Assessment and Risk Management of Toxic Substances*, A Report to the Secretary, Department of Health and Human Services, DHHS Committee to Coordinate Environmental and Related Programmes (CCERP).

Doll, R. (ed) *The Geography of Disease*, *British Medical Bulletin* 40, Churchill Livingstone, Edinburgh.

Farmer, R.D.T. and Miller, D.L. (1977) *Lecture Notes on Community Medicine*, Blackwell, Oxford.

Harvey, B. (ed) (1980–84) *Handbook of Occupational Hygiene*, Kluwer, Brentford, UK.

IARC (1986) *Evaluation of Methods for Assessing Human Health Hazards from Drinking Water*, IARC Internal Technical Report No 86/001, Lyon, France.

Kurihara, M., Aoki, K. and Tominaga, S. (1984) *Cancer Mortality Statistics in the World*, University of Nagoya Press, Nagoya.

Lilienfeld, A.M. and Lilienfeld, D.E. (1980) *Foundations of Epidemiology*, Oxford University Press, Oxford.

Lloyd, O.L.L., Barclay, R., Lloyd. M.M. and Armadale Group Practice (1985) Lung cancer and other health problems in a Scottish industrial town, *Ambio*, **14**, 322–8.

Lloyd, O.L.L., Smith, G., Lloyd, M.M., Holland, Y. and Gailey, F.A.Y. (1985) Raised mortality from lung cancer and high sex ratios of births associated with industrial pollution, *Br. J. Ind. Med.*, **42**, 475–80.

Lloyd, O.L.L., Williams, F.L.R., Berry, W.O. and Florey, C.du V. (eds) (1987) *An Atlas of Mortality*.

Patty, F.A. (ed) (1978) *Industrial Hygiene and Toxicology*, Vol. 1, 3rd edition, Wiley-Interscience, New York.

Royal Society (1983) *Risk Assessment*, Report of a Royal Society Study Group, The Royal Society, 6 Carlton House Terrace, London, SW1Y 5AG.

US EPA (1984) *Risk Assessment and Management: Framework for Decision-Making*, US EPA 600/9-85-002.

WHO (1978) *Principles and Methods for Evaluating the Toxicity of Chemicals. Part 1. Environmental Health Criteria No* 6, World Health Organization, Geneva.

WHO (1986a) *Health and the Environment. EURO Reports and Studies* 100, World Health Organization, Copenhagen.

WHO (1986b) Risk management in chemical safety — conclusions and recommendations, *Sci. Total Environ.*, **51**, 261–3.

WHO (1978–87) *International Programme on Chemical Safety, Environmental Health Criteria*: (1978) 6. Principles and methods for evaluating the toxicity of chemicals, Part 1; (1983) 27. Guidelines on studies in environmental epidemiology; (1984) 30. Principles for evaluating health risks to progeny associated with exposure to chemicals during pregnancy; (1985) 47. Summary report on the evaluation of short term tests for carcinogens (collaborative study on in vitro tests); (1985) 51. Guide to short-term tests for detecting mutagenic and carcinogenic chemicals; (1986) 60. Principles and methods for the assessment of neurotoxicity associated with exposure to chemicals; (1987) 70. Principles for the safety assessment of food additives and contaminants in food.

Chapter 4

Risk Assessment and Management

4.1 INTRODUCTION

Industrialised countries are becoming more and more concerned with the damage on the environment and the side effects on human health resulting from accidents or disasters following the introduction of consumer products, the unfortunate siting of some development projects, or the implementation of hazardous development policies.

Recent examples of environmental and health impacts of consumer products are marine water eutrophication resulting from excessive use of agricultural fertilisers and phosphated detergents; or the reduction from excessive use of freons in domestic vaporisation. Negative environmental and health impacts from hazardous plants such as the Seveso and Bhopal accidents are well known; water projects under warm climates have increased populations of insect vectors of tropical diseases and/or of molluscs as intermediate hosts or agents of parasitic diseases. Development policies such as energy production, transportation or intensive agriculture may have serious environmental and health consequences, for example, in the impact of excessive amounts of fertilisers and pesticides.

Development policies and projects, of course, are needed to ensure the economic and social welfare of the people and cannot be banned unless their negative impacts are immediate and highly visible. The reasonable approach, therefore, is the comprehensive assessment of their potential impact prior to the licensing of consumer products, the construction of development projects, or the approval of a development policy. When the assessment procedure identifies serious environmental and/or health impacts, alternative products, projects or policies should be evaluated or mitigation measures designed and enforced.

Preliminary health and toxicological assessment of pharmaceutical products has been established for a long time in industrialised countries, and now needs to be extended to other products. Similarly, safety assessment of hazardous industrial plants or civil engineering works is a long and well-established practice but now needs to be extended to environmental and health issues. Environmental and health assessment of development policies or the absence of development policies has not been systematically practised in the past, despite environmental disasters such as deforestation, soil erosion and desertification which have resulted from wrong agricultural policies.

This chapter discusses risk assessment in relation to safety in the use of hazardous chemicals and safety considerations in hazardous development projects.

4.2 THE USE OF HAZARDOUS CHEMICALS

Chemicals have become an essential and indispensable part of modern life, sustaining activities and development, preventing and controlling many diseases, and increasing agricultural productivity. However, several of these chemicals may, especially when they are misused, exert adverse effects on human health and the integrity of the environment.

There are at present over 80 000 man-made chemical substances in commercial use throughout the world, of which some 4000, accounting for 99.9% of the total volume, are in common use in many countries. There are about 600 generic pesticides available commercially. In addition, there is a great variety of naturally occurring chemicals used as feed-stocks and many of them are mined or extracted in developing countries. Use of chemicals is essential for economic and social development but a sustainable development process requires that human health and the environment are protected from harmful side effects.

Risks of human and environmental exposure to chemicals, with consequent risk of harmful effects, arise in many ways. In the production, storage, transport, processing, use and disposal of chemicals, such as pesticides in agriculture and the additives in food, human exposure will occur.

Many chemicals will ultimately appear, along with their breakdown products, as pollutants in air, water, food and soil either as residues or wastes. There may be contamination of food by 'natural' toxins such as those produced by fungi. Human exposures can range from the low level over a lifetime to massive exposures during an accident. Many factors affect the outcome, for example, whether the exposure is to a single substance or many substances, the magnitude and duration of the exposure, the physical and biological factors involved and the sensitivity of vulnerable groups in the human population. Consequently, while all countries run the risk of chemical hazards, the conditions and circumstances of exposure and the results will vary widely.

Human health hazards from chemicals thus arise in all areas of socioeconomic development including agriculture, commerce and trade, industry, transportation, recreation and the home. Chemicals found in environmental media such as soil and water can enter the food chain and, under certain circumstances, be involved in endemic diseases or outbreaks of poisoning. Different populations are at different levels of risk. Large segments of populations are exposed to chemical pollutants in the air, water and food, although levels of exposure vary widely. In the working environment, exposure may be high but of limited duration, and affect only selected groups whereas in the general population exposure may be persistently low, but affect the whole population, including its more vulnerable groups such as infants, pregnant women, the elderly, the chronically ill and the malnourished.

As a result of accidents involving chemicals, large groups of people may suddenly be exposed to extremely hazardous situations. In addition to acute effects, there may be delayed and chronic effects, sometimes irreversible, and even affecting the progeny. Exposure to chemicals may contribute to the development of exacerbation of other diseases of multifactorial aetiology.

Rapidly industrialising, developing countries have the full range of problems involving chemicals in the same way as developed countries but have inadequate capabilities to deal with these problems. The least developed countries with non-industrialised economies, which are primarily agriculture but possibly with mining, mineral extraction and some simple formulation of products also face severe problems usually with negligible capabilities for handling them.

Adequate legislation and active chemical safety programmes are needed in these countries to enable them to import safety chemicals and to process, store, transport, handle, use, and dispose of them without risk of adverse effects on human health. Because this preventive approach may on occasions fail or be inadequate, or there may be natural disasters involving the release of chemicals, countries also need the capability for responding to accidents and cases of poisoning and for subsequent clean-up and decontamination. In all cases the specific circumstances of a country must be taken fully into account when considering chemical safety. This applies equally to the manpower training aspects.

The commonly encountered problems concerning chemical safety are:

(i)   lack of knowledge of the chemicals being imported, processed or produced, and used;

(ii)   poor knowledge of the harmful properties of chemicals, how to handle them safely and how to deal with the consequences of misuse. Misuse and abuse of chemicals, not only pesticides, is a major problem;

(iii) where there is information on chemicals, its distribution and use are inadequate;

(iv) absence of regulations and a regulatory infrastructure prevents chemical safety from being implemented;

(v)   shortage of expertise in the scientific, administrative managerial spheres so that governments do not receive the necessary advice on chemical safety matters;

(vi) lack of effective inspection mechanisms and personnel (public health, factory, environment);

(vii) lack of a suitable scientific institutional basis to foster and advance chemical safety;

(viii) lack of awareness and indifference concerning chemical safety on the part of politicians, administrators, managers and the general public.

The growth of chemical industries, in developing as well as developed countries, is predicted to go on increasing for the rest of this century. Chemical safety, that is the control of chemical hazards, is essential if this growth is to be beneficial and not catastrophic for humans and the environment.

4.3 BASIC CONCEPT OF RISK ASSESSMENT AND MANAGEMENT

The policy-decision process to control the risk associated with exposures to hazardous agents is an extremely complex procedure still under development as a scientific discipline. At present the three main steps in the policy-decision process involve research, risk assessment and risk management (Fig. 4.1). This three-step approach has been endorsed by several institutional bodies as the most appropriate process to protect public health from hazardous exposures. However, some parts or intermediate steps of this complex process are widely recognised and frequently used in the policy-decision process.

DOMAIN		RESULTS
Research	Scientists	Scientific Facts
Risk Assessment	(Researchers) Scientists (Assessors)	Scientific Decision
Risk Management	Risk Managers (Politicians, Regulators)	Political Decision

Fig. 4.1. Separation of steps in the risk management process.

- *Research* is a well-defined activity which needs no explanation. It is important to recognise, however, that the data results from research and, as such, are always reproducible.
- *Risk assessment* involves a scientific decision based on the best scientific judgement, and derived from the scientific facts, assumptions, consensus and science policy decisions. Risk assessment, therefore, does not possess the same level of certainty as does scientific research.
- *Risk management* is a political decision and as such it is the responsibility of the whole society represented by politicians and regulators.

Recognition of these three separate steps and understanding of the basic principles involved is critical for further development of the decision-making process.

4.3.1 Risk Assessment
There is no generally agreed definition of risk assessment, but one which reflects the majority of existing descriptions is 'the scientific process of assessing the probability of an adverse effect caused by the exposure to a hazardous substance'.

The recently developed approach to risk assessment shown in Fig. 4.2, is organised to provide two basic answers: how likely is an event to occur; and if it does, how bad could it be in quantitative terms? A similar approach has long been used to estimate risk associated with radioactive exposure and nuclear safety, and some agencies have recently used it to estimate risk associated with toxic chemical exposures.

To develop a comprehensive approach to health hazard evaluation of chemical contaminants in the environment, it is necessary to collect and classify all available and pertinent information on the subject. Sufficient conclusive data are, however, seldom available for most contaminants. In a majority of cases there is a paucity of relevant data, and scientific judgement as well as policy decisions must play a role in establishing acceptable levels of population exposure.

The minimal data base for this purpose should include information on:

- sources of exposure which reviews available monitoring data of present levels of hazardous agents in the environment;
- pharmacokinetic data on the agent, including all available metabolic data in humans and experimental animals; and
- dose-related adverse effects reported in man and other biological systems.

The final output of the risk assessment process should define the values which ideally represents concentrations of chemical compounds in the air or in any other environmental media that would not pose any hazard to the human population. However, the realistic assessment of human health hazards necessitates a distinction between absolute safety and acceptable risk. To aim at achieving absolute safety, would require detailed knowledge of:

- dose-response relationships in individuals in relation to all sources of exposure;
- the types of toxic effect elicited by specific pollutants or their mixtures;
- the existence or non-existence of 'thresholds' for specified toxic effects;
- the significance of interactions and the variation in sensitivity and exposure levels within the human population.

However, such comprehensive and conclusive data on environmental contaminants are not always available. Very often the relevant data are scarce and the quantitative relationships uncertain. Scientific judgement and consensus therefore play an important role in establishing acceptable levels of population exposure. Value judgements are unavoidable, because terms such as 'adverse' and 'sufficient evidence' are not in themselves totally objective, their meaning being based on generally agreed judgements.

Risk assessment consists of one or more objective components such as *exposure assessment, hazard identification* (qualitative assessment) and *dose–exposure assessment* (quantitative assessment).

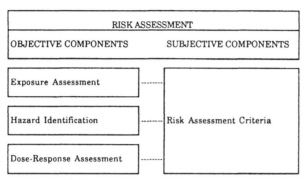

Fig. 4.2. Components of risk assessment.

4.3.2 Exposure Assessment

Exposure assessment has two aspects. First, the general evaluation of actual or anticipated exposure concerning the type, magnitude, time and duration. In general, knowledge of exposure from all sources is needed in recognising the contributions to total body intake from all exposure routes. Additional information relative to special population groups at risk, based on unusual individual susceptibility or unusually high levels of exposure in specific segments of the population must also be given consideration. Most important, the relationship between present levels of exposure to ambient contaminants and the calculated criteria must be carefully evaluated to determine if a human health hazard exists.

The second aspect involves the characterisation of a population, the number of people exposed, a profile of particularly sensitive individuals, and other specific exposure data important for the quantitative estimate of risk for any particular population. This second aspect is usually a part of the risk management process when decisions concern a fixed population.

4.3.3 Hazard Identification

Hazard identification is based on the evaluation of all available data to characterise the strength of evidence, indicating that potential health effects might occur in an exposed human population. The main set of factors that qualitatively and quantitatively influence the risk event and have to be considered are as follows:

- the intrinsic properties of the chemicals;
- the quantity present on the market;
- the modality and plurality of exposure; and
- the size of the population at risk.

Table 4.1
Main physicochemical properties for hazard identification

1. Molecular weight
2. Melting point
3. Boiling point
4. Relative density solids and liquids (D 20/4)
5. Relative density gases (with respect to air)
6. Vapour pressure at 20 °C (Pa)
7. Surface tension (mN/m)
8. Water solubility (g/litre)
9. Fat solubility (g/litre)
10. Flammability
11. Explosive properties
12. Exodising properties

Table 4.2
Main toxicological properties for hazard identification

1. Acute toxicity
2. Skin irritation/corrosion (erythema and/or edema) and/or eye irritataion/corrosion
3. Sensitisation
4. Subacute, subchronic, chronic toxicity
5. Mutagenicity
6. Carcinogenicity
7. Effect on reproduction including teratogenicity

Table 4.3
Main ecotoxicological properties for hazard identification

1. Acute toxicity for fish
2. Acute toxicity for Daphnia
3. Acute toxicity for birds
4. Toxicity for higher plants
5. Effects on algae

The intrinsic properties of a substance are normally defined on the basis of physiochemical, toxicological and ecotoxicological properties, as indicated in Tables 4.1, 4.2 and 4.3. The sum of these three properties determines the intrinsic hazard of a substance, independently from external agents and factors which could influence it.

There is a strict correlation between physicochemical and toxicological or ecotoxicological properties. For example the surface tension is a very important parameter whilst for inhalation and dermal toxicity, lipo-solubility for dermal absorption, hydrosolubility, liposolubility and boiling point play an important role in air and water contamination and so on. So when it is necessary to carry out a sound EHIA of an industrial development project, it is important to first consider the intrinsic properties of chemicals involved in the project.

Physicochemical data and basic toxicological data (acute toxicity, short–medium term toxicity) are available for almost all chemical substances of 'common use' in industrialised countries. Only for a limited part (20%) is it possible to find complete toxicological data (including carcinogenicity studies). In Europe the 79/831 EEC Directive established a completed physicochemical and toxicological testing before the introduction of new chemicals on the market; so in the majority of cases this information is easy to find or to produce, using the standard testing methods proposed by the OECD and accepted by the EEC.

In any case the risk assessment does not limit itself to intrinsic danger, but rather it involves various other factors, such as quantity on the market, plurality of exposure, environmental diffusion, persistence, bioconcentration and size of population at risk.

On the basis of these risk factors, different useful equations have been elaborated in order to obtain both a risk index and a priority assessment in selecting chemicals to be submitted for further studies. One of these studies proposed some general equations which may be used to calculate a risk index (Table 4.4).

Each general equation can be modified on the basis of specific needs; for example if it is necessary to evaluate mutagenicity risk of substances present in all exposure patterns, the equation will be:

Risk index (GE) =

$$(\text{TP mutagenicity}) \times (Q \times BC) \times (PDE + (ED \times P)) \times RP$$

The attribution of weighted coefficients to the additive parameters (PCP, TP, ETP) which contribute to risk potentiality, and to external parameters (multiplicative parameters) (Table 4.5), which may influence such danger multiplying or cancelling it, allow the possibility to express the risk associated with a substance and with some particular conditions with a value.

This methodology may be useful during the preliminary identification phase of health hazard sources in the EHIA of an industrial development project, since it allows the use of a general or specific risk index for each chemical substance involved in the plant processes. Examples are: raw materials, intermediate and final products, their impurities and residues,

Table 4.4

General equations of correlation among different factors influencing risk[a]

Risk index (PDE) = (PCP + TP) × (Q × BC) × (PDE) × RP	(a)
Risk index (EE) = (PCP + TP + ETP) × (Q × BC) × (ED × P) × RP	(b)
Risk index (GE) = (PCP + TP + ETP) × (Q × BC) × ((ED × P) + PDE) × RP	(c)

[a] PDE, plurality of direct exposure (sum of personal exposure, domestic exposure and professional exposure); EE, environmental exposure; GE, general exposure risk index (= PDE + EE); PCP, sum of points ascribed to physicochemical properties; TP, sum of points ascribed to toxicological properties; ETP, sum of points ascribed to ecotoxicological properties; Q, quantity present on the market; BC, bioconcentration; ED, environmental diffusion; P, persistence; RP, size of risk population.

Table 4.5

Multiplicative parameters of intrinsic properties in risk index evaluation

1. Quantity on the market
2. Plurality of direct exposure
   Sphere of personal exposure
   Sphere of domestic exposure
   Sphere of professional exposure
3. Environmental spread
4. Persistence
5. Bioconcentration
6. Size of risk population

substances not directly involved in the process or susceptible to be formed by combustion processes, overheating or probable events.

Differences in risk indexes could be used to determine a rank of possible risks derived by chemicals present in the industrial plant. Table 4.6 presents for some analysed chemicals, a list of risk indexes for environmental exposure obtained by the application of the general equation (b) reported in Table 4.4.

Another important aspect to consider in the hazard identification process, is the distinction between chemical carcinogens and conventional toxicants among substances present in an industrial plant.

Conventional toxicants are characterised by an increase in response frequency and damage seriousness at increasing doses, by the presence of a threshold, reversibility of effects and by a latency period varying in the range of minutes/weeks. Carcinogens, on the other hand, have a response frequency that increases with the dose, but their gravity increase is not

Table 4.6
Risk assessment for environmental exposure[a]

1.	Benzene	2065
2.	DEHP	1656
3.	1,1-Dichloroethane	1650
4.	1,2-Dibromoethane	1552
5.	Aniline	1276
6.	Tetrachloroethylene	1242
7.	Lindane	1176
8.	Parathion	972
9.	Mevinfos	927
10.	Pentachlorophenol	918
11.	1,1,1-Trichloroethane	852
12.	Dichlorodifluoromethane	780
13.	Sulphur trioxide	660
14.	Styrene	436
15.	Formaldehyde	413
16.	1,2-Dichlorobenzene	410
17.	1,4-Dichlorobenzene	394
18.	MPCA	363
19.	1,2-Dichloroethylene	334
20.	Atrazine	319
21.	Acrylonitrile	306
22.	2,2-Dichloroethyl ether	210
23.	Ethylene oxide	200
24.	Trichlorfon	190
25.	Methyl bromide	106
26.	Coumarin	81
27.	Pentane	70
28.	Chloromethyl methyl ether	19
29.	Yellow OB	9
30.	1,5-Naphthylene diisocyanate	0

[a] From Sampaolo and Binetti (1986).

dose-related, they have no threshold, no reversibility to effects, but they do have a latency varying from a few years to ten-year periods. For these reasons it is extremely important in the health component of EIA to consider the possibility of exposure to carcinogens with extreme attention.

These substances generally affect human health through long-term exposures at low doses of contaminants, but the gravity of possible effects places them in the higher range of risk scale.

4.3.4 Dose–Response Assessment

In estimating the health effects on human communities that may result from a development project, it is necessary to compare expected population levels, exposure levels and patterns with existing standards, existing risk evaluations and environmental levels which are reasonably protective of human health (negligible risk for exposed population) at national and international levels.

When information is not available or new important toxicological data has been produced, it may be necessary to estimate or revise these values with a new dose–response assessment.

This kind of information is extremely useful in EHIA for the construction of the 'preliminary hazard identification matrix', for the analysis of the environmental quality of the potentially affected territory and for the assessment of possible adverse effects consequent to possible human scenarios.

In these phases the 'acceptable daily intakes' (ADIs), the 'tolerable daily intakes' (TDIs) and different environmental quality standards (air and water) relative to the examined chemicals represent a useful reference point. When standards are not available, it may be possible to refer to occupational epidemiology and industrial hygiene data using maximum permitted workplace concentrations (MACs and TLVs). These concentrations have to be divided by a factor to include the differences between workplace environment and external environment.

Another important problem to consider in the dose–response assessment is represented by chemical carcinogens, for which a standard definition is not possible from a scientific view point. It may be sufficient for a single molecule to reach the target cell to start the events inducing the appearance of cancer. This is the case of no threshold health effects for which there may be some risk associated with low dose exposure.

In general, dose–response relationships may be obtained by different sources: epidemiological data, toxicity animal experiments and/or short-term mutagenicity tests. Because of the difficulty in defining exposure levels and the possibility to have false negative studies for limited statistical samples in epidemiology and occupational medicine, the experimental studies of animal toxicity are the main source of evaluation.

During the evaluation of toxicological risk from environmental exposure to chemicals, it is important to distinguish between non-stochastic and stochastic effects induced by different mechanisms of action.

The first are typical effects of conventional toxic agents; they are characterised by reversibility, gradation and in particular by threshold effects. This means that toxic agents are characterised by the existence of a No-Observed-Effect-Level (NOEL) below which no adverse effect is observed.

The second are typical effects of genotoxic agents (mutagens and/or carcinogens) which may induce irreversible damage. In this case it is difficult to demonstrate the existence of a 'threshold effect' because, as a general rule, a dose reduction is translated in a reduction of affected population only. On the contrary, the gravity of the damage is dose-independent.

4.3.5 Threshold Effects

In dose–response assessment it is necessary to consider two types of response level:

NOEL: no-observed—(adverse) effect-level, and
LOEL: lowest-observed—(adverse) effect-level

These concepts help to define the threshold region in specific experiments. Generally a free-standing NOEL has little utility if there is no indication of its proximity to the LOEL, because the NOEL may be many orders of magnitude below the threshold region (Federal Register, 1980).

The NOEL and LOEL identification are highly related to the number of experimental animals used for the treatment groups (a limited number of animals may show a dose with effect as NOEL) and to the extension of the observation period.

When these doses are identified, an appropriate uncertainty factor (safety factor) has to be applied to NOEL to consider the uncertainty in extrapolating animal data to man.

The safety factor value is normally chosen on the basis of nature of toxic effects and dose–response curve, size and type of population to be protected, and quality of toxicological information.

In some cases regulatory agencies have adopted general guidelines in establishing uncertainty factors. EPA criteria reported here may be a useful example (Federal Register, 1980):

Uncertainty factor 10: Valid experimental results from studies on prolonged ingestion by man, with no indication of carcinogenicity.

Uncertainty factor 100: Experimental results of studies of human ingestion not available or scanty (e.g. acute exposure only) with valid results of long-term feeding studies on experimental animals, or in the absence of human studies, valid animal studies on one or more species. No indication of carcinogenicity.

RISK MANAGEMENT	
OBJECTIVE COMPONENTS	SUBJECTIVE COMPONENTS
Risk Assessment	Risk Perception
Cost Impact	Political & Social Constraints
Cost benefit Analysis	Intangible Values

Fig. 4.3. Components of risk management.

Uncertainty factor 1000: No long-term or acute human data. Scanty results on experimental animals with no indication of carcinogenicity.

The result of the ratio NOEL/SF is a daily dose or a total daily exposure level which can be assumed for a lifetime with negligible risk for human health. It may be expressed as ADI or TDI. In the case of air and water quality standards, on the other hand, the contribution of air and water matrices to the total contamination has to be considered.

When starting from animal experiments for carcinogenic risk assessment it is necessary to extrapolate results from animals to humans, considering differences in body surface and in time of exposure between species. A useful concept is that of the equivalent dose for man (ED_{man}) which may be calculated by multiplying the animal dose by a species factor and a time factor as indicated below:

$$ED_{man} = D_{animal} \times \text{Species factor} \times \text{Time factor}$$

where

$$\text{Species factor} = \frac{\text{Body surface (man)}}{\text{Body surface (animal)}}$$

$$= \frac{(\text{man body weight})^{\frac{2}{3}}}{(\text{animal body weight})^{\frac{2}{3}}}$$

These quantitative risk estimates should not be regarded as being equivalent to true cancer risk, but rather rough estimates of risk which may be a useful basis for policy-makers to establish a degree of urgency and to set priorities on public health problems.

4.3.6 Risk Assessment Policy

Risk assessment must often rely upon insufficient or inadequate scientific data. Assumptions, therefore, are needed in each phase of the risk assessment process to fill the various gaps in the available scientific knowledge. In dealing with the uncertainties associated with the mixture of scientific facts and assumptions, it is important to have a firm risk assessment policy in order to avoid inconsistencies in the final decisions, or to resolve points of current controversy in the scientific community. These policy statements should be changed from time to time when new scientific evidence appears.

4.3.7 Risk Management

There is no generally agreed definition of *risk management*, but the term is used to indicate the decision-making process which combines the risk assessment results with the socio-economic, technical, political and other considerations to reach a conclusion on how to control and manage exposure to suspected hazardous agents. Risk management as shown in Fig. 4.3 is composed of objective and subjective components. Objective components are risk assessment, cost impact and cost–benefit analysis. *Risk assessment* expressed as a risk value associated with the statistical measure of uncertainty is numerical information as a result of the risk assessment process. *Cost impact* establishes the cost of various options and determines differences among them.

The critical element of this process is the differential *cost–benefit analysis*, which starts from the determined health (or other) risks, and then, through the evaluation of associated economic costs with control measures and available technologies, and with the weighing alternatives, selects the most appropriate public health action. The most common subjective components of the risk management process are risk perception, social and political constraints, and intangible values.

4.3.8 Risk Perception

Risk perception is the most serious obstacle of the risk management. A rational decision is usually biased by a risk perception, because in certain cases perceived risk by the public differs from the true risk estimated on the basis of the best available scientific knowledge. As a general rule, the public tends to overestimate low probability/high consequence events, such as nuclear power, and underestimate high probability/low consequence events, for instance smoking. Because cancer often appears many years after the smoker starts to smoke, and because the smoker is usually the only victim, smoking continues to be practised with little or no restriction. Contrary, despite its reasonable safety record (even with the Chernobyl accident insight), restrictions are imposed on nuclear industry by regulations, standards and costly intervention.

4.3.9 Social and Political Constraints

Social and political constraints are additional limitations placed on the objective risk management components. Cultural background, as well as the special groups interest are for the policy-makers of equal importance as the objective components.

4.3.10 Intangible Values

Intangible values are usually the most significant although the least quantifiable areas of the risk management. The clear view of a mountain peak, or the beauty of a natural forest are not quantifiable in economical terms, and neither are risks to life or limb or an unborn child or a 70-year-old man or woman, and these values cannot be determined solely on the basis of cost–benefit analysis. Therefore, the importance of this value to society is a matter of opinion and must be determined by the particular community.

4.4 SOME GENERAL CONCLUSIONS

The approach outlined provides a way of separating the scientific facts from the scientific assumptions and political issues. Separating the risk assessment from the risk management process ensures that assumptions critical in dealing with scientific uncertainty in the risk assessment process are not biased by an eventual desired risk management outcome. It also necessitates a re-evaluation of many processes practised by some international organisations.

Political decisions on the acceptability or otherwise of a certain risk should be made by national authorities within the framework of a risk management process. However, due to differences in available economic resources some countries may be less stringent in applying regulatory measures. Finally, it is for the people of each country to choose which kind of life (or risk) they want to accept. This simple, but important fact is not generally recognised by many policy-makers and/or programme managers. In some international programmes, there is a tendency to impose 'uniform standards' valid for the entire world. This attitude comes from the time when scientists were recommending standards based on arbitrary decisions.

Some organisations, such as the European Community, are able to make risk management decisions, which are mandatory for all Member States. These political decisions are usually made after long negotiations between those concerned, and when resources are available to compensate countries for economic and other differences.

Exceptions relate to decisions on risk management dealing with transboundary pollution, where international coordination in the standard setting is not only justifiable, but highly desirable.

International organisations and agencies may assist governments in developing a risk assessment for different hazards. The risk assessment process is still in transition or evaluation, and is often a mixture of scientific facts, assumptions, consensus and science policy decisions.

There is also a tendency to confuse risk assessment with scientific research. Research deals with the scientific facts, and risk assessment is the scientific decision based on the best scientific judgement. It may be factual and scientifically robust, or it may be no more than an educated guess. Assumptions are often needed in each phase of the assessment process to fill the various gaps in the scientific knowledge in order to complete the risk assessment. Dealing with future events and predicting situations many years ahead involves many uncertainties, consequently, recommendations concerning risk assessment may conclude that 'because of the lack of appropriate epidemiological data, no firm recommendations can be made'.

4.5 RISK ASSESSMENT AND MANAGEMENT OF DEVELOPMENT PROJECTS

4.5.1 Introduction
Risk assessment and environmental impact assessment of development projects have developed as separate research fields, and with practitioners from different disciplines. However, both are methods of predicting the effects of policy decisions. EIA has tended to focus upon the identification of impacts whereas risk assessment involves the rigorous analysis of the probability of that impact occurring and the magnitude of effect.

A number of major accidents in recent years have focused public attention on industrial plant location and plant safety. They are:

(i)   Flixborough chemical plant explosion 1977.
(ii) Seveso toxic cloud release 1976.
(iii) Three Mile Island nuclear power plant meltdown 1979.
(iv) The release of toxic methyl isocyanate from a chemical works in Bhopal 1984.
(v)   Chernobyl power plant explosion 1986.

These and other accidents have led regulatory authorities, industry and the public to pay greater attention to plant location and safety. The assessment of risk, in many cases, is now a requirement at the planning or pre-operational stages of development projects.

4.5.2 Risk and Hazard

A *hazard* may be defined as the inherent property of a system to cause damage, and *risk* is the likelihood of that harm being realised and is therefore a measure of probability. Risk assessments have traditionally focused on impacts upon public health and effects measured in terms of morbidity, mortality or premature death.

It is unusual for other impacts upon the environment to be expressed in probabilistic terms although some risk assessments have considered other effects, for example, ecological considerations.

4.5.3 Methods

4.5.3.1 Deterministic risk assessment

Risks with an engineering content may be estimated on the basis of case histories, such as statistics of routine traffic accidents or failures in box girder bridges. Where this body of data exists, comparisons may be made between new proposals and existing case histories. This deterministic approach incorporates implicit value judgements as to what is an acceptable standard of practice.

However, in view of the potential for major consequences involved in engineering failures, it is unacceptable to wait for disasters to occur in order to build up a body of case histories as a basis for policy decisions.

In these circumstances, the use of probabilistic risk assessment, involving the description of hazards in terms of the risks of failure and the magnitude of impacts has been advocated.

4.5.3.2 Probabilistic risk assessment

Probabilistic risk assessment consists essentially of three distinct steps:

(i)   Identification of the hazards (circumstances having the potential to cause harm).
(ii) Identification of initiating events that might lead, via various accident scenarios, to the risk manifested.
(iii) Quantification to calculate the probability of occurrence and the associated consequences of the event.

  The identification of hazards will depend upon the process involved. In the case of a nuclear fuel reprocessing plant, a hazard may be the uncontrolled release of fission products into either the atmosphere or the sea. By studying plant design and processes, a series of failures can be identified that would be required to bring about such a 'top event'. This is commonly undertaken by means of *fault trees* (Fig. 4.4).

An example is the probability of loss of offsite power or mains water supply necessary for the cooling of high active liquor storage tanks in a nuclear fuel reprocessing plant. From probability theory the frequency of the top event can be calculated.

Following a systems failure, numerous mathematical techniques exist to estimate impacts upon human health, agriculture and many other aspects of the environment. For example, the gaussian plume theory may be used to estimate the dispersion of an atmospheric release of radioactive gaseous material. Modelling procedures exist to estimate the uptake of these substances by man and to calculate effects upon health, and upon agriculture.

4.5.4 Expression of Risk

Risk may be defined as the *product* of frequency and magnitude of consequence. The expression of that risk is important when ranking alternatives, for example, the death rate *per passenger mile* for car travel is about twice that for air travel; by expressing risk in terms of *passenger hours* air travel is seen to present twice the risk of that by car.

Four principal categories of risk expression have been identified as follows:

- Individual risk
- Death per unit measure of activity, e.g. man years of employment
- Loss of life-expectancy
- Frequency versus consequence lines for risk to society (fc lines)

Of these individual risk and societal risk (fc lines) are commonly used in the assessment of major hazards, e.g. nuclear, chemical and petrochemical plants.

Individual Risk (IR) is commonly used as an expression of mortality data, in terms of the fractional death rate over time in a given population. For example, if in any year, 5000 people are expected to die in road accidents in the United Kingdom and if the whole population (50 million) is assumed to be uniformly at risk, then IR = 5000/50 million or 10^{-4}. The assumption of uniformity is often false, sub-groups of a population may be preferentially exposed, e.g. cyclists or motorcyclists. Therefore an expression of IR will include contributions from sub-groups at high risk and from groups effectively unexposed.

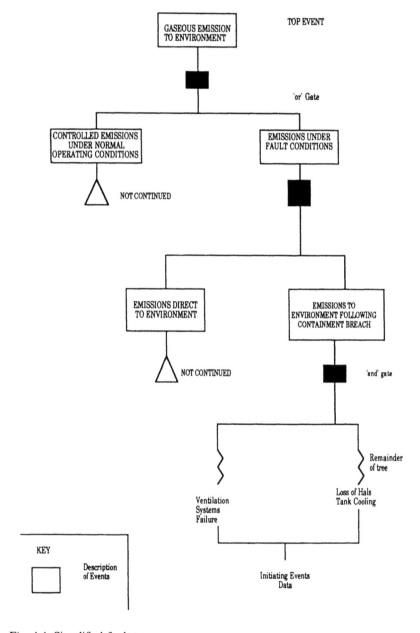

Fig. 4.4. Simplified fault tree.

In the assessment of major hazards, IR is the product of the frequency of occurrence of a particular accident in relation to the probability of it affecting an individual. For example, with the example of the nuclear fuel reprocessing plant, the top event identified of an atmospheric release from a high active liquor tank may have a frequency of 10^{-4}. By modelling the outcomes of that release, taking into account a range of meteorological conditions, evacuation procedures and dose–response relationships, the level of harm experienced by an individual at a given location can be expressed in terms of probability of harm over time. The calculation may be varied to consider preferentially exposed groups or alternatively IR may be summed to produce an average individual risk expression.

Societal risk does not focus on the probability of harm to individuals, rather upon the total consequences of an event, i.e. the total number of people likely to be affected and the probability with which that event will occur.

4.5.5 Risk Acceptability

Society in its perception of risk, places greater importance on large consequence events such as a single event causing 5000 deaths than it does to that of 5000 accidents per year due to traffic accidents. This relationship between magnitude of consequence and frequency is implicit in the expression of societal risk and fn lines shown in Fig. 4.5. By calculating the probability of various accident scenarios versus the probable outcome, a series of frequency (f) versus outcome (n) relationships having N number of casualties can be established.

The question of what level of risk is acceptable to an individual or to society is central to the issue of risk assessment. It has been suggested that: 'An acceptable risk is a risk where probability of occurrence is so small, whose consequences are so slight, or whose benefits (perceived or real) are so great that a person, group or society is willing to take that risk'.

Rarely however, is the issue that straightforward. In particular the decision-maker has to decide if a level is acceptable, and take account of the distribution of risks and benefits and the possibility of compensation.

Various schemes have been devised, considering levels of acceptability for early and delayed deaths for both individual and societal risk. Kinchin (1978) proposes individual risk criterion for nuclear reactors at 10^{-6} per year for early death and 3×10^{-5} for delayed deaths. Other criteria are proposed by the Royal Society, who state for IR that 10^{-6} or possibly 10^{-7} are appropriate, but did not define societal risk criteria.

In the Netherlands, considerable effort has been directed to defining

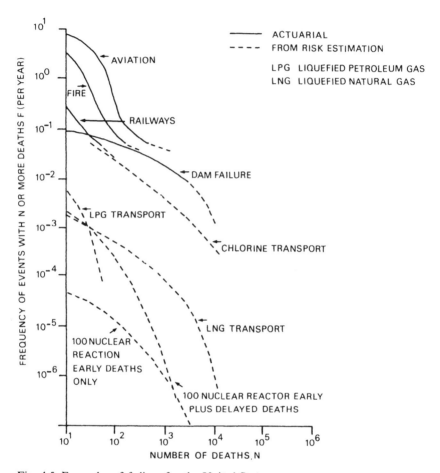

Fig. 4.5. Examples of *fn* lines for the United States.

levels of acceptable societal risk. The Provinciale Waterstaat Gröningen has developed comprehensive guidelines to be used in assessing major hazards. The Gröningen scheme (Fig. 4.6) goes further than previous schemes, in that it treats casualties as equivalent fatalities, such that serious injury or death is weighted as 1, where as broken limbs are allocated a weighting 0.01 The scheme does not cater for improbable high consequence events, in that an upper limit of 1000 'equivalent deaths' is the maximum permissible, even at the most extreme 'probability'. The Gröningen scheme is probably the most stringent of those proposed on this subject.

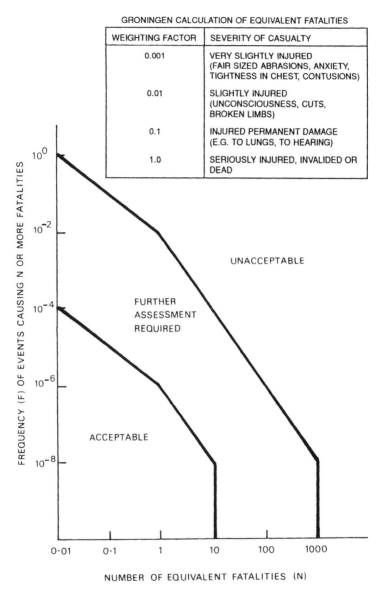

Fig. 4.6. The Gröningen criteria.

4.6 RISK ASSESSMENT IN EIA

It is unusual to find the concept of probability embodied within environmental assessments. Invariably, EIAs focus upon deterministic studies and upon impacts which are certain or simply fail to recognise 'uncertainty' in those that are not. Uncertainty, and lack of scientific data are particularly important when modelling the consequences of accidents or discharges.

There is a need for a careful definition of both risk events and of the resources at risk. Developers and objectors should attempt to seek agreement on the definition of risk events, prior to the preparation of an environmental impact statement.

The use of quantification in the prediction of impacts within EIA is to be welcomed and encouraged. However, the explicit recognition of uncertainty presents the author and reviewer of an EIA with a dilemma, for whilst the scientific community applaud attempts to qualify statements and accept that rarely are events certain, the public do seek certainty and the taking of the right decision.

Although there rarely is a requirement that risk should be embodied in an EIA (indeed such a requirement is absent from the terms of the EEC Directive), practitioners are recognising that the treatment of risk in the assessment process can be beneficial. However, there is clearly a long way to go before risk assessment becomes an integral part of all environmental impact statements.

4.6.1 The Management of Environmental Risks

In recent years, considerable effort has been placed in developing the tools for the management of environmental risks. The major thrust has been towards improving the methodology, whereby the risk of detrimental changes in the environment caused by specific human action can be anticipated, and their extent and probability estimated. Principles for the management of environmental risk have been developed via environmental impact assessment (EIA) whereby developers are required to document the environmental consequences as part of the normal process of the granting of permits. EIA has been based on the need to understand the regional and local consequences resulting from the siting of chemical industrial production and storage facilities; by the power industry; and in the development of transport systems for the movement of chemicals and raw materials especially liquid fuels. The general approach has been to determine the direct environmental consequences of the initial installation, the anticipated effects of normal operations, and the potential consequences of irregular operations and catastrophic failures both within and without the fence-line. The formalisms of such risk assessment are based on inventories of failure mode-consequence calculations including estimates of probabilities of chains of events and outcomes.

Because of the potential for enormous direct losses in capital investments resulting from plant or transport failure, and the direct costs of cleanup and secondary costs of liability accompanying catastrophic chemical or radiation release or plant or transport failure, the driving force for EIA is economic. The economic importance of loss-protection (e.g. public, corporate and private liability) has resulted in the development of a separate industry devoted to the various aspects of quantitative, probabilistic and comparative risk assessment and risk management.

The human health consequences of commercial activities are of no less importance than the environmental consequences. In particular many environmental changes brought about by industrial activity have direct and indirect health impact. The recognition of this fact has led to the development of a range of tools for environmental health risk management, similar to those for environmental risk management.

4.6.2 Environmental and Health Impact Assessment of Development Policies

Major development policy decisions can have direct and indirect consequences on the environment and on public health. In many cases the direct environmental and health impacts of development policy can be determined by examination of the type of projects or activities envisaged to be controlled or encouraged by the policy decision, together with a review of existing EIA and EHIA exercises. Frequently, the health and environmental impacts are subtle and are the result of unintended or unforeseen changes which accompany the promotion or restriction of development. One such area of concern is the health impact of population shifts through the creation or loss of markets or workplaces, changes in emigration policy, and adjustments of market or trade barriers. These impacts can be due to shifts, either for the better or worse, of the quality of and ease of access to primary health care, changes in lifestyle and social status brought about by urbanisation, etc.

One major policy decision illustrative of the scope of the problem is the unification of the European Economic Community by 1992. Implementation of this decision will be accompanied by a wide range of increased opportunities for lifestyle shifts of a large part of the population. Changes in the availability of health care and health insurance, and in access to medical procedures and diagnostic tools will alter base-line data on which national or local health planning is dependent. Increased availability of financial and monetary products and instruments (pension schemes, loans, investment opportunities) may result in population shifts (especially for certain high risk or high medical cost age groups) which will in turn be accompanied by changes in demand and efficiency of local medical facilities.

Planning policy decisions can be a long time in development, but the response of market forces is extremely rapid. A systematic analysis of the potential for health impact of policy decisions will enable adequate preparation to be made both locally, nationally and regionally.

4.6.3 Environmental and Health Impact Assessment of Development Projects

Methodologies have been developed for EIA of major development projects, especially with respect to prediction of changes in base-line concentrations of priority substances. Similar methodologies have enabled the health impact of such projects to be determined, based on traditional concepts of chemical risk. In addition, a range of new ideas have been introduced in terms of what constitutes 'health impact' including questions of shifts in risk distribution, equity, and other social phenomena which have public health consequences. On the other hand, philosophical shifts in legislative attitudes have resulted in changes in technical approaches to health impact assessment, especially through the replacement of point estimates of risk (frequently based on worst case assumptions): the conservative approach by risk ranges accompanied by detailed discussion of uncertainties. Improvement in EHIA of development projects will come through improvements in the quality of the technical data base for effect estimation, as well as through changes in the formalism and structure of the conceptual frameworks used. Towards this end, consideration could be given to the development of a unified framework for EHIA of development projects on a regional and local level. This would ensure that the same inventory of risks was considered regardless of locality, while variations in local priorities could be accommodated in EHIA activities.

4.6.4 Methodology

The methodologies used for the environmental impact assessment of development projects, are also required to screen those environmental impacts with health significance and, to assess the significance of human exposure to these impacts.

The following approaches are proposed:

Step 1: screening and scoping of primary environmental impact (tools — checklists)

Step 2: identification and quantification of secondary or tertiary impacts (tools — networks, energy diagrams)

Step 3: screening of impact with health significance (tools — public health knowledge, environmental health data)

Step 4: human exposure assessment, identification and quantification of population exposed to the environmental hazards (tools — geographical census, knowledge on ecotoxicological pathways of exposure)

Step 5: identification and quantification of risk groups exposed (tools — population survey, medical knowledge)

Step 6: computation of resulting mortality and morbidity (tools — human exposure assessment, dose–response curves, epidemiological)

Step 7: human risk management procedure

Step 8: comparison of alternatives or development of mitigation measures

Step 9: risk management decision

4.6.5 Practical Examples of Health Impact Assessment of Development Projects

A number of practical examples of the health impact of different development projects are given in Annex B. The development projects include:

- Coal-fired power station
- Water reservoir construction
- Comparison of alternative electricity production policies based on solid fossil fuels and on the nuclear cycle
- Agricultural policy of putting under cultivation, forest or range lands in tropical countries, and intensive agriculture in industrial countries
- Urban transportation policy
  The examples identify primary impacts, risk groups and mitigation measures.

4.7 PUBLIC PARTICIPATION IN RISK ASSESSMENT

Increasing public awareness of technological risk has been accompanied by increasing public participation in the decision-making framework.

Over the last 20 years the public have become more aware and willing to be involved in public inquiries and hearings. Members of the public participate directly, or through membership of associations, amenity and pressure groups. Some groups may be opposed to a specific project expressing the 'not in my back yard' philosophy, others under the green umbrella focus upon more generic environmental issues.

Public safety is commonly the key issue in major development inquiries, however, the complexity of the issue, the adversarial nature of inquiry proceedings and statutory regulations all mitigate against effective participation. For example, the reporter at the public inquiry held onto the application for planning permission for a Nuclear Fuel Reprocessing Plant at Dounreay, Scotland, ruled that there was no need for a full safety analysis to be presented at the inquiry as the issue of plant safety would be dealt with by the regulatory authority, the Nuclear Installations Inspectorate (NII). All stages of site licensing, as operated by the NII, however, are undertaken privately between themselves and the developers with no scope for public involvement.

Effective public participation measures are essential if controversial projects are to be sited without direct action against them.

REFERENCES AND BIBLIOGRAPHY

Andersen, M.E. et al. (1987) Physiologically based pharmacokinetics and the risk assessment process for methylene chloride, *Toxicol. Appl. Pharmacol.*, **87**, 185–205.

Anderson, M.W. et al. (1980) A general scheme for the incorporation of pharmacokinetics in low dose risk estimation for chemical carcinogenesis: example vinyl chloride, *Toxicol. Appl. Pharmacol.*, **55**, 154.

Andrews, R. (1986) Environmental impact and risk assessment: learning from each other. In Miller et al. (1986).

Astill, B.C. (1983) Structure–activity relationships in priority setting of chemicals. General considerations, report prepared for OECD.

Beanlands, G. (1986) Risk in EIA: the perspective of the Canadian Ferdera; Environmental Assessment and Review Office. In Miller et al. (1986).

Bidwell, R. et al. (1987) Public perceptions and scientific uncertainty: the management of risky decisions, *EIA Rev.*, **7.1**, 3–23.

Caldwell, L.K. et al. (1982) *A Study of Ways to Improve the Scientific Content and Methodology of Environmental Impact Analysis*, Indiana University, Bloomington, IN, OSF Research Grant No PRA – 79–10014.

Clarke, H. and Kelly, G.N. (1981) *MARC — The NRPB Methodology for Assessing Radiological Consequences of Accidental Releases of Activity: NRPB — R127*, HMSO, London.

Coulston, F. and Pocchiari, F. (1983) *Accidental Exposure to Dioxins: Human Health Aspects*, Academic Press, New York.

Edwards, A.W. (1976) *Likelihood*, Cambridge University Press.

Federal Register (1980) EPA-water quality criteria documents, *US Federal Register*, **45**(231), 79318.

Federal Register (1984) Chemical carcinogens: notice of review of the science, *US Federal Register*, **49**(100), 21594.

Federal Register (1979) Scientific basis for the identification of potential carcinogens and estimation of risks, *US Federal Register*, **44**(131), 39858.

Food Safety Council (1980) Quantitative risk assessment, *Food Cosmet. Toxicol.*, **18**(6), 711.

Gehring, P.J. et al. (1979) Risk of angiosarcoma in workers exposed to vinyl chloride as predicted from studies in rats, *Toxicol. Appl. Pharmacol.*, **49**, 15.

Gifford, F.A. (1981) Estimating ground-level concentration patterns from isolated air pollution sources: a brief summary, *Environ. Res.*, **25**, 126.

Giroult, E. (1987) Introduction to environmental health risk management, *Proceedings of the Course on the Quantitative Assessment of Toxicological Risk for Human Health*, ISTISAN Report 87/50, pp. 3–66.

Griffiths, R. (1986) *Risk Expressions and Risk Criteria*, Evidence to EDRP Public Inquiry, available from Scottish Office, Edinburgh.

Grima, A.P. et al. (1986) *Risk Management and EIA: Research Needs and Opportunities*, CEARC, Quebec, Canada.

Hasset et al. (1980) Sorption properties of sediment and energy related pollutants, EPA – 600/3–80–41.

IARC, (1982) *IARC Monographs*, Suppl. 4.

Irving Sax N. (1979) *Dangerous Properties of Industrial Materials*, Van Nostrand Reinhold, New York.

Jarret, D.E. (1968) Derivation of British explosive safety distances, *Ann. NY Acad Sci.*, **152**, Art. 1.

Kenaga, E.E. and Goring, C.A. (1980) Relationships between water solubility, soil sorption, octanol–water partitioning and concentration of chemicals in biota. In *Aquatic Toxicology* (J.G. Eaton et al. eds), ASTM STP 707, Amer. Soc. Testing and Materials, Philadelphia, PA, p. 78.

Lipnick, R.L. and Dunn, W.J. (1982) *Proc. 4th European Symp. Chemical Structure–Biological Activity, Quantitative Approaches*, Bath, UK.

McCall, P.J. et al. (1983) Estimation of environmental partitioning of organic chemicals in model ecosystems, *Residue Rev.*, **85**, 231.

Mackay, D. (1979) Finding fugacity feasible, *Environ. Sci. Technol.*, **13**(10), 1218.

Mackay, D. and Paterson, S. (1982) Fugacity revisited, *Environ. Sci. Technol.*, **16**(12), 654.

Magill, P.L. et al. (1956) *Air Pollution Handbook.*, McGraw-Hill, New York.

Mill, T. (1981) Minimum data needed to estimate environmental fate and effects for hazard ranking of synthetic chemicals, *Proc. Workshop on the Control of Existing Chemicals (OECD)*, Berlin, p. 207.

Miller, C.T., Kleindorfer, P.R. and Munn, R.E. (1986) *Conceptual Friends and Implications for Risk Research*, IIASA, Austria.

Neely, W.B. et al. (1974) Partition coefficients to measure bioconcentration potential of organic chemicals in fish, *Environ. Sci. Technol.*, **8**, 13.

OECD (1983) *Part 1: Physical/Chemical parameters and Biodegradation; Part 2: Toxicity and other Biological Tests*, Annex VII to Env/Chem/LD/83.15.

Richet, C. (1983) Sur le rapport entre la toxicite et les proprietes physiques des corps, *C.R. Soc. Biol.*, **54**, 775.

Ramsay, C. (1985) Hazard and risk, *Proceedings of the 1985 Seminar on EIA*, University of Aberdeen, CEMP.

Royal Society (1983) *Risk Assessment a Study Group Report*, The Royal Society, London.

Sadee, C. et al. (1976) The characteristics of the explosion of cyclohexane at the Nipro (UK) Flixborough Plant on 1st June 1974, *J. Occup. Accidents*, 203–5.

Sampaolo, A. and Binetti, R. (1986) Elaboration of a practical method for priority selection and risk assessment among existing chemicals. , *Regul. Toxicol. Pharmacol.*, **6**, 129–54.

Shaper, K.J. and Seydel, J.K. (1982) Quantitative structure–pharmacokinetic relationships and drug design, *Pharmacol. Ther.*, **15**, 131.

Strauss, W. (1971) *Air Pollution Control*, Wiley-Interscience, New York.

Swann, R.L. et al. (1982) A rapid method for the estimation of environmental parameters K^{ow}, K^{oc} and water solubility, *Residue Rev.*, **85**, 17.

Suter, G.W. et al. (1986) Treatment of risk in EIA. In Miller et al. (1986).

WHO (1976) *Environmental Health Criteria 1: Mercury*, WHO, Geneva.

WHO (1978) *Environmental Health Criteria 6: Principles and Methods for Evaluating the Toxicity of Chemicals. Part 1.*, WHO, Geneva.

WHO (1982) *Rapid Assessment of Sources of Air, Water and Land Pollution*, WHO Offset Publication No 62, Geneva.

WHO (1982) *Safe Processing and Ultimate Disposal of Redundant Products Containing or Emitting Potentially Toxic Chemicals*, WHO Regional Office for Europe, Copenhagen.

WHO (1982) *Health Hazards from Mercury in the Mediterranean Region*, ICP/RCE 221(1)/7.

WHO (1982) Risk assessment, *Health Aspects of Chemical Safety Series*, Interim Document 6.

WHO (1987) Health and safety component of environmental impact assessment, *Environmental Health Series*, No 15, WHO, Copenhagen.

Zapponi, G.A. and Bucchi, A.R. (1987) Iclorometano: analisi matematico-statistica del rischio cancerogeno legato a particolari usi della sostanza (caffe decaffeinato e lacche per capelli), Rapporto ISTISAN 87/51, p. 50.

Zapponi, G.A. (1988a) Methods for the health component of EIA of industrial development projects, *IX International Seminar on Environmental Health Impact Assessment*.

Zapponi et al. (1988b) Reproducibility of low dose extrapolation comparison of estimates obtained using different rodent species and strains, *Biomed. Environ. Sci.*, **1** (in press).

Chapter 5

Environmental and Public Health Impact Assessment

5.1 INTRODUCTION

Health factors have often received inadequate attention during the formulation of development policies and planning of projects. In particular, the health component of EIA is weak and needs to be strengthened through the use of environmental health impact assessment (EHIA). This aims first to identify and predict the impacts of a development on environmental parameters that have a strong health significance, in other words environmental health factors. Then, using this and other information from epidemiological, toxicological and risk assessment studies, an attempt is made to identify, predict and evaluate potential changes in health which may be caused by a particular development proposal.

5.2 DEFINITIONS

Methods for EHIA are mechanisms which have been devised to assist in the collection and organisation of information such as checklists and matrices. The EHIA process itself is based on that shown in Fig. 5.1. The EHIA activities can be subdivided into a number of broad categories similar to those in the EIA process itself. They are:

- *Impact identification* which refers to the determination of those impacts requiring investigation. In practice, the lack of knowledge on the effects of different developments on a variety of environments makes this a complex task;
- *Impact measurement* which refers to a quantitative estimation of magnitude, such as noise levels at a housing area 2 km from a

Step 1	Assessment of primary impacts on environmental parameters
Step 2	Assessment of secondary and tertiary impacts on environmental parameters
Step 3	Screening of impacted environmental parameters for health significance (identification of environmental health factors). Preliminary identification of environmental health impacts
Step 4	Prediction of how project will affect exposure of populations to environmental health factors
Step 5	Prediction of how project will affect size of health risk groups
Step 6	Computation of predicted health impacts in terms of mortality and morbidity, if possible
Step 7	Definition of significance and acceptability of adverse health impacts
Step 8	Identification of mitigation measures to prevent or reduce significant adverse health impacts
Step 9	Final decision on whether or not the project should proceed

Fig. 5.1. The EHIA process. Based on Giroult (1984).

proposed industrial plant. Other aspects of impacts may be measured, such as the geographical extent of the probability of an impact occurring.

- *Impact interpretation* refers to the need to determine the importance of an impact. For example, how important is a 5 dB(A) increase in noise levels for the inhabitants of a particular housing area. The relative importance of impacts in comparison with other impacts of a different nature may be considered under impact 'interpretation'.

- *Impact communication to information users* refers to the presentation of information to help decision-makers and interested public to come to some conclusions on the merits or demerits of a proposed project.

- *Impact monitoring and auditing* refers to the enforcement of standards and identifying changes in environmental parameters due to development activities and/or operations.

Not all methods are suitable for all of these activities; some can assist with only one activity, others with several. There is also a subtle distinction between 'methods' and 'techniques'. In general the term 'methods' applies to the identification, interpretation, communication and mitigation/

monitoring aspects of the process, while 'techniques' refers mainly to measurement and prediction of specific impacts. The emphasis here is on methods.

Source materials are particularly important if the health component of EIA is to be improved. One of the major obstacles to the development of EHIA has been the EHIA 'information gap'. Although environmental health impacts are characterised by special difficulties — for example, they are frequently indirect, non-specific, long-term, and difficult to quantify — a great deal of factual information on specific impacts is available. Unfortunately, it is often difficult to identify, locate and access such information when required. Furthermore, few EIA and EHIA practitioners have detailed medical knowledge of the full range of environmental health impacts, so summary information is needed as to the nature of environmental health impacts, including their causes and effects. Improvements in environmental health status, especially in developing countries, could probably be achieved simply by improving access to existing information and providing information summaries for non-experts. Example of a valuable source material for EHIA is the WHO Environmental Health Criteria series (WHO, various dates).

A number of other key terms require definition. *Risk* is 'a compound measure of the probability and magnitude of adverse effect' (Lowrance, 1980). *Risk assessment* thus has two facets: it includes both risk of accident (a probability) and risk of morbidity or mortality (adverse health effects of variable magnitude). Risk assessment is a sub-component of the broader EHIA process. It is concerned primarily with impact measurement and prediction. Risk assessment techniques will not be covered in detail in this paper.

Hazard is 'the inherent property of a system that could cause injury or damage'. A useful, comprehensive classification of hazards is as follows:

- infectious and degenerative diseases
- 'natural' catastrophes
- failure of large-scale technological systems
- discrete, small-scale technological systems
- low-level, delayed-effect hazards
- sociopolitical disruptions
  These hazards fall within the scope of EHIA, but the focus here is on the assessment of environmental health hazards in the community, rather than in the workplace (occupational health hazards).

Epidemiology and toxicology are tools for risk assessment. *Epidemiology* is the science of estimating risk from human population samples, by analysing how morbidity and mortality are distributed in relation to age, sex, spatial and other factors. *Toxicology* is the study of the undesirable biological actions of chemical substances. It attempts to estimate the impacts of such substances on humans, animals and the environment. A commonly used toxicological technique is to study the effects of the substances on animals in the laboratory in order to predict the effects on humans by extrapolation.

5.3 LIMITS AND CONSTRAINTS ON THE EHIA PROCESS

Some of the more significant limits and constraints on the EHIA process are summarised in Fig. 5.2.

5.3.1 The Nature of Environmental Health Impacts
A number of limits and constraints on the EHIA process arise from the nature of environmental health impacts themselves. Environmental health impacts tend to be complex. Environmental systems consist of complex inter-relationships of linked components, and the effects of changes in environmental systems on health are complicated further by the variety of exposure pathways and human sensitivities. This means that it is often difficult to identify and predict links between environmental change and human health.

Many environmental health impacts are indirect secondary and tertiary effects. For example, inadequate sewage treatment may lead to groundwater contamination, which may in turn cause gastrointestinal disease. Furthermore, effects are frequently non-specific, that is, the same effect may be caused by different factors. For example, lung cancer may be due to one of a number of different carcinogens. Environmental health factors can also interact. Environmental chemicals may interact with each other to give a reductive effect, or an additive effect, or a synergistic effect (where the combined effects arise from the influence of nutritional, dietary and other lifestyle factors such as smoking and alcohol intake. Nutritional factors are of particular importance in developing countries, where susceptibility to disease is often increased by malnutrition. In developed countries, the influence of lifestyle factors is now a major health concern.

The spatial and temporal characteristics of environmental health impacts also give rise to difficulties in the EHIA process. In order to estimate the effects of a chemical that has been released to the atmosphere, for example, it is necessary to know both its environmental concentration and its environmental distribution. Sometimes impacts can occur at considerable distances from the source, as in the case of acid rain. Impacts can also occur in different time spans. Some environmental health impacts, such as noise impacts, manifest themselves immediately; others, such as the toxic effects of bioaccumulation of heavy metals, may only appear after a long period of time has elapsed. In addition, environmental health impacts are dynamic, they can change over time and some effects are reversible, some irreversible.

Probability may be a serious problem. Results of epidemiological and toxicological studies are expressed in terms of statistical probability. It is therefore often impossible to be precise in the identification, prediction and assessment of environmental health impacts. This in turn makes decision-making difficult.

In addition, impacts are subject to problems of perception. The socio-political dimension of environmental impacts has been discussed in Chapter 2. Environmental health impacts are especially emotive, and may be interpreted differently by different individuals and social groups. Risk assessment professionals and others vigorously debate the relative emphasis that should be given to calculated and perceived measures of risk. The most appropriate conclusion seems to be that although every effort should be made to make EHIA as 'objective' and scientific as possible in identification and prediction of impacts, the role of value judgements in the process should not be ignored.

5.3.2 Limits of Scientific Knowledge

There are a number of limits in the application of scientific knowledge to EHIA. First, there have been problems in the identification hazards, which often owes much to chance. For example, the association of cigarette smoking with bronchitis and heart disease was discovered early this century, but it was not until 1962 that statistical evidence of a causal relationship between cigarette smoking and lung cancer was produced. For quite a number of other suspected health hazards it has proved difficult to establish, unequivocally, an environmental health risk.

A second, related constraint arises from the limitations of epidemiology methodology, such as the inability to control exposure levels and population mobility, and the difficulty in determining history of previous exposure and other relevant variables. For effective human epidemiological studies, adequate data resources are necessary, preferably in mutually compatible, collated and accessible form. Unfortunately, very few countries have developed such data resources.

In addition, there are inadequacies in biological knowledge of chemical

```
•  The nature of environmental health impacts
   -  many indirect secondary and tertiary effects
   -  impacts frequently non-specific
   -  interactions between environmental health factors common
   -  complex spatial and temporal characteristics
   -  highly probabilistic character of many environmental health impacts
   -  perception problems

•  Limits of scientific knowledge
   -  problems in identification of hazards
   -  limitations of epidemiology methodology
   -  inadequacies in biological knowledge of:
      chemical toxicity and environmental disease processes;
      mechanisms for biological defence and activation;
      biological receptors which act as targets for toxic reactions
   -  limitations of analytical methodology

•  Biological variation
   -  heterogeneous human populations
   -  difficulties with animal toxicological studies including different
      disease susceptibilities, metabolic differences and problems with
      extrapolation models

•  Resource constraints
   -  manpower: training and personnel deficiencies, poor intersectoral
      coordination
   -  finance
   -  information
```

Fig. 5.2. Summary of limits and constraints on the EHIA process.

toxicity and environmental disease processes, mechanisms for biological defence and activation and biological receptors which act as targets for toxic reactions.

Finally, analytical methodology is limited. Assessment of environmental risks to health are only as reliable as the methods used, so the specificity, sensitivity and reproducibility of these methods need to be carefully determined and defined.

5.3.3 Biological Variation

Constraints are also imposed on the EHIA process by biological variation. The problems of biological variation affect both major avenues of approach to environmental health risk assessment. The epidemiological approach is made difficult by heterogeneity of the population, with many highly susceptible sub-groups such as the old, the young, the malnourished and those with inherited disease. In the experimental animal approach, there is the problem of how animal toxicological data may be used to predict risk to humans. Assumptions about the relative susceptibilities of humans and animal species — for example, the common assumption that humans are as sensitive as the most sensitive animal species tested — are unreliable. There are also limitations caused by metabolic differences among species. Choice of appropriate mathematical models for extrapolation of dose–response relationships is another area of difficulty.

Studies in vitro of human cell cultures have been used in an attempt to overcome the difficulties caused by differences between experimental animals and humans, but these too have limitations, related to the absence of systematic functions which would exist in an intact animal.

5.3.4 Resource Constraints

Resources for EHIA and risk assessment are an area of constraint, but also of opportunity, for by establishing clear priorities and directions it may be possible to make better use of limited resources.

The development of EHIA has in the past been constrained by the lack of appropriately trained and experienced staff, but EHIA training and the inclusion of health professionals in a multidisciplinary EIA team are now being actively encouraged by WHO and other national and international agencies. Related to the lack of appropriate manpower are organisational difficulties such as poor coordination between government authorities with environmental health responsibilities, including health authorities, economic and physical planning departments, and pollution control regulatory authorities. This is another area requiring improvement.

Careful use of financial resources for EHIA is important. The thousands of new man-made chemicals produced every year, make the scientific assessment of risk of environmental chemicals very expensive. There is a need to prioritise the assessment of those chemicals to which the degree of population exposure will be greatest. The scope of assessments should also be defined at the outset, in respect of the risk groups to be given greatest attention. Moreover, there is a need to improve information feedback from human observations after exposure to environmental health factors, and to develop and coordinate epidemiological data banks, as mentioned earlier.

5.4 METHODS FOR EHIA

The development of specific methods for EHIA is a comparatively new phenomenon. The efforts of the public health professions have been directed largely at the assessment and improvement of current health status, representing a curative approach. In recent years, health impact assessment (HIA) has developed. HIA is a preventive approach, which aims to predict the direct effects of a development on human health, in practice, for two main reasons. First, it is difficult, given the uncertainty of prediction, to derive precise figures for changes in morbidity or mortality arising from a development. Second, given this level of uncertainty, such figures are likely to be highly controversial, and political sensitivity may prevent their publication outside of confidential internal documents. Moreover, as EIA is increasingly used as an aid to planning and decision-making, it seems wise to integrate the health component into the overall EIA process, through EHIA.

By predicting future changes in environmental health factors (environmental parameters having a strong health significance), it is possible to broadly indicate the potential changes in health which may be caused by a development. These indications can then be used to make changes in project design and/or siting, to prevent or mitigate any adverse environmental health impacts. They can also be used by local experts to plan health care programmes.

EHIA methods, as stated earlier, are tools to assist in the EHIA process. Referring again to Fig. 5.1, it can be seen that Steps 1 and 2 in the EHIA process are the assessment of primary, secondary and tertiary impacts on environmental parameters. This is in effect the normal EIA process. EIA methods are documented elsewhere and will not be covered here. However, the information derived from the EIA process forms an important input to Step 3 of the EHIA process, the identification of environmental health factors. All three initial steps involve 'screening' and 'scoping'.

Methods for identification of environmental health factors in Step 3 are based on epidemiological and toxicological evidence of causal links between environmental parameters and health effects. Step 4, the prediction of changes in exposure to environmental health factors, involves the study of exposure pathways. Steps 5 and 6 use epidemiological and toxicological information on dose–incidence and dose–response relationships between environmental parameters and health effects to predict changes in risk due to exposure to environmental health factors, and to calculate likely health impacts in terms of morbidity and mortality, if possible. In Step 7, the significance and acceptability of adverse health impacts are evaluated and compared with the beneficial impacts of the proposed development. Step 8 is the identification of measures to prevent or mitigate significant adverse health impacts and Step 9 is decision-making.

In the following review of existing EHIA methods, no attempt has been made to classify the methods because they vary in several dimensions. Some apply only to certain types of development project or disease agent; some are of use only at certain stages in the EHIA process; and a number of generically different mechanisms are utilised. The more 'general' methods that apply to a wide range of types of development project are discussed first. Methods for assessment of health effects of environmental pollution (physical and chemical factors) are then examined, and finally, methods for assessment of impacts due to communicable and vector-borne disease (biological factors). The advantages and disadvantages for each method are outlined. The methods are not necessarily intended to be applied as they stand, but they do provide a source of ideas on how to approach the EHIA of a particular project. They may be adapted, combined and developed to suit individual situations.

5.4.1 Rapid Assessment

Rapid assessment is a method, developed by WHO (1982), which can be used for EHIA of existing pollution situations, where previous industrial and urban development may have been subject to inadequate controls, or where the combined effect of many small pollution sources has created serious problems. It can also be used to give a preliminary analysis of the pollutant loading likely to be caused by new industrial development. It is primarily a tool for measurement and prediction of changes in exposure to environmental pollutants. It is of less use for EHIA of water development projects, where the major environmental health impacts will be communicable diseases.

Rapid assessment was devised with the needs of developing countries in mind. Environmental pollution control programmes are desperately needed in many developing countries, but in these countries there is often a serious lack of information as to the size and scale of health hazards, because limited resources are available for environmental monitoring and control. The rapid assessment method is relatively simple and applicable in most countries. The aim is to produce, in 6–8 weeks, a preliminary assessment of the most urgent pollution control problems in a given area, and those which may be emerging. Figure 5.3 shows the steps in the method.

The method is based on available information such as industrial production figures, fuel usage, number of motor cars, number of houses connected to sewers, various statistics on population, and the like. These are used as indicators of consumption and output for industrial and urban processes. They are multiplied by waste load factors (derived from industrial and urban processes) to provide estimates of pollutant emissions. For instance, if the annual fuel consumption in tonnes of an oil-fired power plant is known, and the SO_2 emitted from combustion of a tonne of oil is also known, an estimate of the SO_2 emitted annually by the plant is obtained by multiplying the two figures. An example of the form in which results appear is given in Fig. 5.4. The value of such information by itself may be quite limited, but it can be useful when analysed in the context of other local information, which is also assembled in the course of the study.

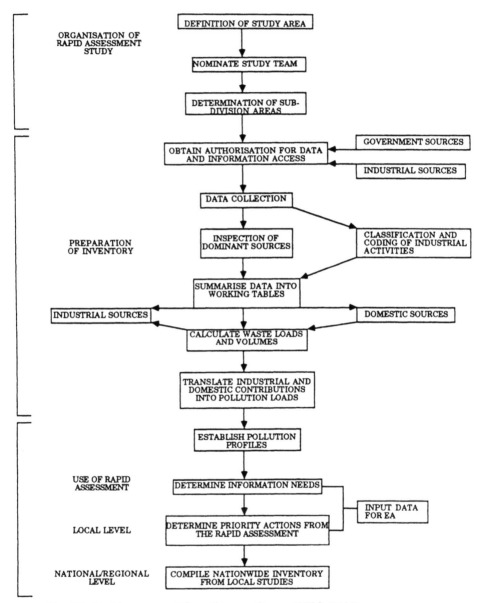

Fig. 5.3. Rapid assessment flow diagram. Source WHO (1983).

	Particulate matter		SO2		NOx		HC		CO	
	tn/y	Contri-bution	tn/y	Contri-bution	tn/y	Contri-bution	tn/y	Contri-bution	tn/y	Contri-bution
Power plant	309	1.3%	10645	53.3%	3923	48.5%	77	2.2%	125	0.3%
Industry	21492	96.0%	7100	35.6%	1465	18.2%	755	22.1%	119	0.2%
Road traffic	454	2.0%	1920	9.6%	2175	27.0%	2072	60.4%	44807	97.9%
Port activity	101	0.5%	262	1.3%	305	3.8%	200	5.8%	280	0.6%
Airport	13	0.1%	18	0.1%	139	1.7%	319	9.3%	460	1.0%
Domestic sources (excl. wood burning)	29	0.1%	21	0.1%	36	0.5%	5	0.2%	6	0.0%
Total	22398		19966		8043		3428		45797	

Fig. 5.4. An example of the form in which the results of rapid assessment are presented.

For example for assessment of the air pollution situation, details of meteorological conditions and spatial distribution of pollution sources are important. The results of the assessment are used alongside other information such as local health statistics and regional development planning information to set priorities for the establishment of pollution control and prevention strategies.

The method is not without problems, particularly of data collection and interpretation. Data may have to be extracted from unpublished sources by means of factory visits. Screening is necessary to determine the significance of data, and data reliability may be poor. Many assumptions are required in the course of the study, in order to deal with data deficiencies. Waste load factors derived from other parts of the world may have to be used, and may not necessarily be appropriate. Finally, it is not clear how the results are related to local health statistics to predict changes in risk and morbidity and mortality effects due to changes in exposures to environmental pollutants. More work is needed in these areas, and the method needs to be tested and developed in practice. Even so, it may represent an advance on current approaches to assessment of environmental pollution in developing countries.

5.5 METHODS FOR EHIA OF WATER DEVELOPMENT PROJECTS

As a WHO priority area, the development of methods for EHIA of water development projects have received considerable attention. Two important studies document the methods for EHIA of water development projects. They are described below.

5.5.1 Panel of Experts on Environmental Management for Vector Control (PEEM) (WHO/FAO/UNEP)

The third meeting of PEEM in 1983 discussed 'Methods of forecasting the vector-borne disease implications in the development of the different types of water resources projects'. The types of project covered were irrigation and drainage, rural water supply, hydro-electric schemes, flood control and other water-related developments. Although most major water developments have been the subject of detailed prior study, secondary effects including ecological and health effects have frequently been neglected, giving rise to serious unforeseen adverse impacts.

The Panel reviewed methods for ecological prediction of the influence of water projects on vectors of disease, and suggested a method for forecasting vector populations. Three tables present indications of the growth and decline of different disease vector populations as influenced by a range of activity components related to dam construction. Decreases in vector populations are indicated by a minus sign ($-$), increases by a plus sign ($+$) and no change by a zero (0). One table covers the project construction stage (see Fig. 5.5) another the project operational stage, and a third deals with long-term effects. The method, a type of matrix,performs three functions:

(i)   It can help to define the relationships between the various construction-related activities and vector populations under study.
(ii) It can assist workers in considering the development process in a logical, sequential and functional framework.
(iii) It can help to predict the changes likely to take place with various inputs.

Additional information requirements of the method include entomological data, and a wide range of other planning information such as population characteristics, environmental hygiene patterns, bio-epidemiological aspects, climatological details, topographical description, operational engineering information and details of local flora, fauna and agriculture. These are relevant because of the great variety of factors affecting disease transmission and exposure. There is also a checklist of engineering design details, which need to be considered along with information given in the matrix in the study of any particular project.

Overall, the method appears to be one of the best tools yet developed for identification and measurement of environmental health impacts related to water development projects. Its advantages include: coverage of the different development stages; systematic examination of the relationships between construction-related activities and vector populations; and provision of a considerable amount of information on the nature of the impacts concerned. A limitation is that it applies only to dam construction projects under given conditions (listed in the report) which will obviously not be fulfilled in all cases.

The Panel also examined epidemiological considerations for water development projects, and suggested methods for identifying exposed populations and risk groups and quantifying the risks to which they will be exposed. This is a particularly valuable approach, which could usefully be adapted and applied to EHIA of other types of development project.

A type of health assessment matrix (Fig. 5.6) can be used at various stages in any water development project to evaluate the health status of the population, and to derive the disease potential. The two dimensions used in the matrix are population categories and environmental health risks. Environmental health impacts are identified and their magnitude is estimated on a 1–5 scale, where 1 is the least magnitude and 5 the greatest. The magnitude scores are recorded in the matrix, providing an easily understood visual summary. The matrix is intended to furnish a basis for analysis of changes in and needs of affected populations and for subsequent epidemiological surveillance. This method is unusual because it is concerned not only with the identification of environmental health impacts, but with the issues of exposure and risk groups, which generally need to receive greater consideration in the EHIA process. It is also good in that it includes the time element and mechanisms for measurement and interpretation of impacts, through impact scores.

5.5.2 Environmental Health Impact Assessment of Irrigated Agricultural Development Projects
The study was prepared by Environmental Resources Ltd (ERL) for WHO Regional Office for Europe (1983). It contains a 'complete' set of methods for EHIA of irrigated agricultural projects, with four main sections covering:

(i)   identification of potential impacts
(ii) prediction of impacts on environmental health
(iii) mitigating measures
(iv) organisation and presentation of information for the decision-maker

It provides a useful consolidation and structuring of existing information

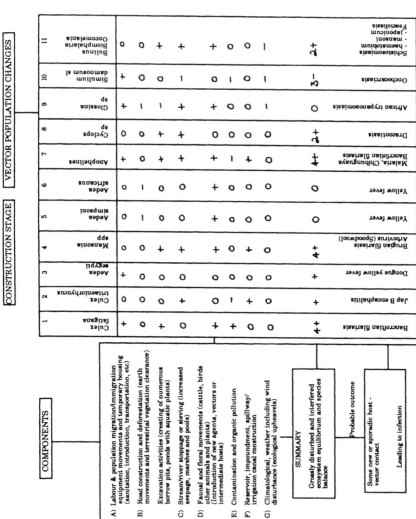

Fig. 5.5. Effects of a water development project on vector populations in the construction stage. Key: 0 = no change in the vector population; − = decrease in the vector population; + = increase in the vector population. Note: all + and − symbols are only qualitative expressions of change. Source: PEEM Secretariat (1983).

Fig. 5.6. Matrix for assessment of environmental health risks on different population categories. Source: PEEM Secretariat (1983).

A Direct Impacts on people in the project area

- Will new diseases or new strains of the disease be introduced by immigrations of construction workers or new settlers? Will these affect new settlers or residents or both?

- Will relocated communities be exposed to diseases to which they have little or no immunity?

- Will new settlers be exposed to locally endemic diseases to which they have little or no immunity?

- Will food, waste or water cycles aggravate sanitation and disease problems?

- Will housing and sanitary facilities become overburdened, misused or not used at all, leading to conditions conducive to increases in water washed diseases and spread of communicable diseases by the faecal-oral route?

- Will soil and water be contaminated by excreta, facilitating spread of communicable disease?

- Will introduction of migrant workers cause increases in venereal disease among workers and subsequently residents?

- Will new settlers and relocated communities be exposed to physical, social and cultural changes leading to psychological strains and traumas? These may include changes in lifestyle and employment.

- Will changes in food supplies lead to possibilities of malnutrition, nutritional deficiencies or toxic effects? These effects may occur because of:

  - introduction of western-style convenience foods;
  - changes in staple foods - possibly using unfamiliar toxic plants as substitutes for usual foods;
  - contamination of soil or agricultural water supplies with toxic substances;
  - reduced productivity of soils caused by hydrological changes (waterlogging etc), mineralisation or pollution of ground and surface waters;
  - reduced productivity of fisheries caused by hydrological changes or water pollution;
  - change in availability of trace metals in soils caused by hydrological changes (lowering or raising of water table etc)

- Will effluents and emissions, or substance released intentionally into the environment (eg. pesticides) pollute air or water or soil presenting a threat to human health?

- Will irrigation of fields increase opportunities for human contact with water borne, water based and water related disease?

- Will traffic in the area, and therefore road accidents, increase as a result of the development?

- Will new industries and similar activities attracted to the area by growth, result in pollution of air, soil or water or noise, with subsequent impacts on human health?

B INDIRECT IMPACTS THROUGH EFFECTS ON DISEASE VECTORS

- Will new vectors be introduced into the area from upstream as a result of hydrological changes?

- Will new vectors be introduced into the area on vehicles, animals, transplanted plants, soil, etc?

- Will existing vectors be infected or reinfected by contact with infected humans coming into the area?

- Will the prevalence and distribution of existing infected vectors be changed by changes in the availability of suitable habitats for breeding and survival? These changes may result from hydrological changes (water velocities, temperature, depth, standing water, etc), morphological changes (bank slopes, cover, etc), climate changes (rainfall, humidity) and biological changes (vegetation, predators, etc). They may affect presently infected or uninfected areas.

C DIRECT IMPACTS ON WORKERS

- Will migrant workers be exposed to locally endemic diseases to which they have little or no immunity?

- Will migrant workers be exposed to psychological strains and traumas from changes in living and working conditions?

- Will workers be exposed to physical threats to their safety (injuries, deaths) or chemical and physical hazards to health (toxic substances, noise, vibrations, radiation, high pressures, etc)?

- Will workers be particularly exposed to contact with water and thus with water associated disease during their work?

- Will workers be exposed to dangerous animals during their work (snakes, scorpions, etc)?

- Will adequate supplies of food be provided to prevent malnutrition and minimise spread of disease (eg. by use of itinerant food vendors)?

D IMPACT ON HEALTH SERVICES

- Will health and other social services be overburdened with consequent effects on health of residents and workers?

Source: various publications including

World Bank; "Environmental, Health and Human Ecologic Considerations in Economic Development Projects". World Bank Source: Various publications including Washington, 1974 (New edition in press)

Fig. 5.7. Questionnaire checklist of potential health impacts of water resource developments and irrigation projects. Source: WHO Regional Office for Europe (1983).

on environmental health impacts of irrigated agriculture. The material is clearly assembled and presented, and the recommended EHIA process is explained step by step. The main features of the study are described below. More information will be found in the study report.

First, the study provides a helpful classification of water-related diseases according to their means of transmission. In the section on identification of potential impacts, two checklists are presented. One is a simple checklist of possible components of water resource developments and irrigation projects, the other a questionnaire checklist of potential health impacts (Fig. 5.7).

Prediction of impacts on environmental health is subdivided into five steps:

(i)   obtaining baseline information of the environment and human health
(ii) predicting changes to environment and habitat factors favouring disease transmission
(iii) predicting effects of changes in these factors on disease vectors
(iv) predicting changes in human exposure to disease
(v)   predicting future incidence of disease

Guidance is given on the information required for each of these steps. The guidance is mainly in the form of checklists of information requirements. There is less guidance on how to analyse the information collected. Particular difficulties are encountered with step (v): the authors conclude that formal methods are not readily available for predicting direct health impacts, that is, future incidence of disease. Expert judgement is recommended to relate information on future changes in the environment and in human exposure to disease incidence.

The section on mitigating measures is comprehensive. Types of mitigating measures available are classified into three groups:

(i)   environmental modification, that is large-scale alterations to the form of the environment, such as clearance of land before a project commences, drainage and dewatering of areas round a project
(ii) environmental manipulation, that is, small-scale control of the environment during the operational phase using physical, chemical and biological methods
(iii) modification or manipulation of human behaviour or habitats to reduce man–vector–pathogen contact.

Mitigation measures available for different disease vectors are summarised in Fig. 5.8, and further details are given in the text. Another table summarises mitigation measures to be considered at different stages of a development.

Finally, methods for organisation and presentation of information to decision-makers are examined.

The chief advantage of the study is its comprehensiveness. For the first time, specific guidelines for EHIA of an important type of development project have been developed and summarised in a single document. While there remains some scope for further development of methods contained in the study — especially those for prediction of disease exposure and disease incidence — the general approach is very sound.

5.5.3 Disease Vectors and Parasite Reservoirs

Water developments can substantially affect environmental health by introducing new disease vectors and parasite reservoirs. The creation of open water promotes the growth of aquatic and terrestrial vegetation, alters ground water levels and attracts animals or birds. An assessment must first identify the principal species of vectors, animals and plants which will be encouraged or discouraged by the project. It must be assumed that if a vector occurs in the region, and a favourable breeding site is created by the project, then the vector will colonise the site.

5.5.4 Human Exposure

The significance of a vector community is determined by its contact with human beings. The degree of contact can be influenced by altering the population size or behaviour of the vector or adjusting the lifestyle of the affected population. Each species of vector has a preferred time and location for feeding; for example, the risk of tse-tse fly attack is greatest in woodland during the day, whereas certain mosquito species are more dangerous at night and indoors. Equally, environmental and behavioural changes may be appropriate to limit contact with unsafe water. An EIA would be required to predict both the changes in vectors and other environmental health factors and the likely exposure of local populations to them.

A method developed by the Panel of Experts on Environmental Management for Vector Control (PEEM) (PEEM Secretariat, 1983), can assist in the systematic consideration of relevant factors. The method uses tables to indicate the growth and decline of different disease vector populations as influenced by the different proposed project activities (see for example, Fig. 5.9). Decreases in vector populations can be indicated by a minus sign, increases by a plus sign, and no change by a zero. The sum

Vector or intermediate host	Diseases transmitted	Environmental modification						Environmental manipulation						Modification or manipulation of human habitation or behaviour				
		Drainage (all types)	Total earth filling	Deepening and filling	Land grading	Velocity alteration	Impoundment	Clearing and burning of terrestrial vegetation	Shading or exposure to sunlight	Water level fluctuation	Sluicing for flushing	Aquatic vegetation control	Salinity regulation	Water supply and sewerage	Screening and bednets	Refuse collection and disposal	Land use restriction	Improved housing
Anopheles mosquitos	Malaria	++	++	++	++	+	−	+	+	+	+	+	++	+	+	+	+	+
Aquatic snails	Schistosomiasis	++	+	++	+	+	−	−	−	+	+	+	+	++	−	−	+	−
Culex and Aedes mosquitos	Filariasis, viral and other diseases	++	+	+	+	+	−	+	+	+	+	+	+	+	+	+	+	+
Blackflies	Onchocerciasis	−	−	−	−	+	++	−	−	−	+	−	−	+	−	−	+	−
Houseflies	Infantile diarrhoea	−	−	−	−	−	−	−	−	−	−	−	−	++	+	++	−	+
Tsetse flies	African trypanosomiasis	−	−	−	−	−	−	++	−	−	−	−	−	−	−	−	+	−
Triatomid bugs	Chagas' disease	−	−	−	−	−	−	+	−	−	−	−	−	−	+	−	−	++
Rat fleas	Plague	−	−	−	−	−	−	−	−	−	−	−	−	−	−	++	−	++
Cyclops	Dracontiasis	−	−	−	−	−	−	−	−	−	−	−	−	++	−	−	−	−

Fig. 5.8. Types of mitigation measures available for control of vector-borne diseases. Key: − = little or no directly demonstrated value, or not applicable; + = partially effective (some species); + + = partially effective (most species). Small dams — adverse effect; large dams — good effect. Source: World Health Organization (1980) *Environmental Management for Vector Control; 3rd Report of the WHO Expert Committee* WHO Technical Report Series No 649, WHO, Geneva.

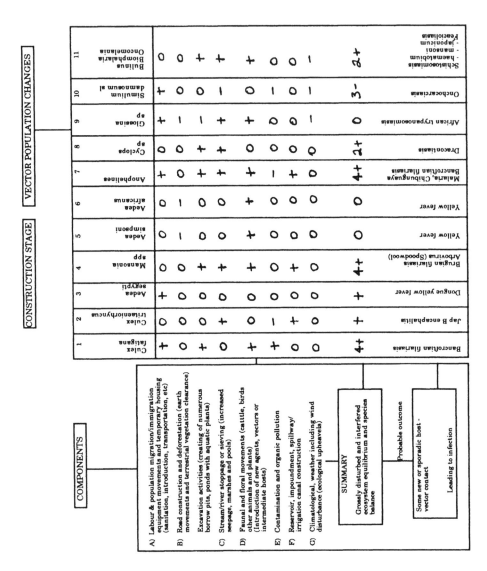

Fig. 5.9. Effects of a water resource development project on vector populations in the construction stage. Key: 0 = no change in the vector population; − = decrease in the vector population; + = increase in the vector population. Note all + and − symbols are only qualitative expressions of change. Source: PEEM Secretariat (1983).

of these changes can provide an indication of the likely overall change in vector population size. Separate tables can be prepared for different stages of a project, e.g. construction and operational stages. The method, a type of matrix, can perform four functions. It can:

(i) help to define the relationship between the various construction and operational activities and vector populations;
(ii) help to predict likely changes in vector populations over time;
(iii) be used to demonstrate the effects of changes in project design on vector populations;
(iv) provide a general indication of the magnitude and significance of changes in vector populations.

Although the example shown applies to a specific type of project under given environmental conditions, the same approach can be adapted for use for other projects and conditions.

5.5.5 Population Studies in Relation to Water Resource Developments

Population studies in relation to water resource developments are also concerned with the identification of exposed populations, possible exposure pathways, and risk groups. The most important questions relate to:

- human use of water following the development; whether water within the irrigation system will be used for washing, drinking, fisheries etc.;
- location of human settlements; whether these will be close to water bodies, spillways and in general to vector and disease organism breeding and feeding areas;
- levels of bodily contact with water; for example through washing in water and water-based recreation; through labour in inundated fields, through maintenance of the irrigation system;
- human personal hygiene and eating habits; waste disposal, future hygiene practices, eating habits related to disease (for example, eating raw fish or plants which may be infected by parasites);
- immigration to the area; how the migrants will settle and when, information on social habits and health status on potential migrants.

Prediction of human contact with water-related diseases is highly complex, not only because of the need for detailed information on social attitudes and lifestyles, but also because it is likely that lifestyles themselves may change significantly following a major water resource development. For example, development of a reservoir may allow fishing to become important; irrigated agriculture may attract a new population of farmers. Figure 5.10 shows an example of the kind of method which can assist in the identification of exposure pathways and risk groups. It is a matrix which enables different population categories to be systematically related to different health risk pathways. Potential health impacts can thus be identified, and if wished, their magnitude can be estimated on a 1–5 scale. The matrix can be used as a basis for detailed exposure assessment and for subsequent epidemiological surveillance. It can be prepared for different project phases to help predict changes in health impacts through time.

5.6 METHODS FOR PREDICTING HEALTH EFFECTS

5.6.1 Identification of Health Effects

During the 15 years in which EIAs have been produced, a number of methods for impact identification have been proposed and used, some of which may have utility for the identification of health effects of a project.

The first decision which has to be made is whether the project appears prima facie to have significant environmental and health effects and ought to be subject to an EIA. The term which is used to describe the activity of selecting projects for assessment is 'screening'. Screening may be facilitated if there is a legislative or regulatory requirement to prepare assessment for certain project types. The EC Directive 6553/85, for example, introduced mandatory assessments for, among other things, 'integrated chemical installations'. Discretion may be exercised in determining if a particular plant comes within such a broad definition. Other forms of listing may provide limits on the productive or storage capacity of facilities, or may list particular processes which are considered hazardous. In the absence of precise guidance on the need for an EIA, the screening process may involve meetings between the proponents, regulatory agencies and other interested parties, which may or may not be open to the public. The nature of the project and the anticipated impacts will indicate to the authorities whether or not an assessment is required. Public perception of potential impacts may be such that an assessment is undertaken against the views of expert opinion.

A related activity to screening is the selection of significant impacts to be assessed, that is, the scoping of the study. Again, scoping is usually a process of meetings between interested parties and experts, taking into account public concerns. Details of the processes and materials involved in the project and knowledge of similar facilities would suggest environmental health factors and issues to be included in the assessment. The involvement of environmental health professionals at this stage is crucial.

Among the methods for impact identification, 'checklists' are the

POPULATION \ RISKS	Transmissible by Direct Contact	Preventible through vaccination	Sexually transmitted	Acute respiratory	Transmitted by sylvatic vectors	Transmitted by urban or domiciliary vectors	Zoonoses	Water or food transmission	Malnutrition	Toxic, through contamination	Occupational	Traumatic	Mental	Chronic
Existing Urban														
Existing Rural														
Fishermen and Families														
Project workers														
Organised workers														
Spontaneous Services														
Marginal resettled														
Marginal displaced														
Marginal newly formed														
Visitors and Tourists														
Downstream														

Fig. 5.10. Identification of exposure pathways and risk groups. Source: PEEM Secretariat (1983).

simplest. These ensure that systematic and comprehensive consideration is given to potential effects, but their weakness is that they do not identify secondary or combined effects. Modifications have produced questionnaire checklists and scaled or weighted checklists, which involve rather more assessment of the significance of the impacts. The value of checklists is that they can be designed for a range of project types and may be applied by relatively unsophisticated assessment teams (for example the US Aid questionnaire checklist and the World Bank simple checklist).

Matrices have been used in EIAs to identify impacts arising from each development action. A list of proposed actions is displayed on the horizontal axis, while environmental components are listed on the vertical axis. The matrix is 'completed' when all possible effects are marked at the appropriate intersection. This approach does not facilitate consideration of indirect, combined and cumulative effects because each impact is linked separately to the causal development activity. The strength of matrices is that by combining two checklists, that of the development and of the environmental elements, all simple direct impacts should be identified. An advantage of matrices is that they provide a clear presentation of predicted impacts.

Another method which has been introduced to assessment studies is overlay mapping in which the spatial distribution of environmental features and predicted impacts are successively superimposed. Overlays may make a useful contribution to the prediction of environmental health impacts. Maps of the predicted distribution of pollutants could be combined with population distributions and maps of sensitive locations, such as schools and hospitals, to identify where health effects are likely to be significant. They could also indicate areas which would be subject to the combined effects of more than one substance.

Systems analysis approaches have also been used to identify impacts. Networks, for example, attempt to integrate impact causes and consequences through identifying inter-relationships between actions and environmental elements, including those representing indirect effects. An alternative is to attempt to model the energy flows within a system and to identify and assess changes in the energy system which would result from a development. Man is part of a very complex system, and network methods may contribute to the understanding of the relationship between a project and environmental health. Systems analysis is already a powerful tool for ecotoxicologists and toxicologists studying the behaviour of substances in the environment. Network methods may allow integration of impact assessment and toxicological approaches.

5.6.2 Estimation of Health Effects

After the potential health effects of a development proposal have been identified, the significance of the effects must be assessed. Estimation of health effects requires knowledge of baseline situations and the predicted impacts. Baseline information consists not only of population data, particularly on high risk groups, but includes details of any special pre-development health characteristics of the population.

In the case of industrial projects, predicted impacts include the effects of routine releases of toxic substances and the risk of accidents involving explosion, fire or leaks of hazardous materials. The estimation of health effects on human communities requires knowledge of basic principles of toxicology. Consequently, in this step it is advisable to involve public health authorities and toxicology experts.

The first and simplest analysis to be carried out is comparison of expected pollution levels, exposure levels and patterns with the existing standards, such as acceptable daily intakes (ADIs), threshold limit values (TLVs) and tolerable levels, if available. All these parameters are generally the result of a chemical risk assessment procedure, in which all the relevant health aspects should have been considered.

Many areas of the world have air and water quality standards, established by agencies with regulatory authority to control environmental degradation. In addition, international agencies also propose environmental criteria. In so far as these standards are based on the prevention of negative health impacts from environmental pollutants, they may be used as short-cuts in interpreting the significance of health effects. If the concentration of a substance is within the relevant standard, it presumably will not cause health problems. Approached in this way, environmental criteria appear to simplify greatly the impact interpretation of health effects.

5.7 MITIGATING MEASURES

An important purpose of impact assessment is to identify ways in which the negative effects of a proposal could be prevented or minimised, by altering aspects of the development or introducing protective measures in the environment. The mitigating of deleterious health effects is a function of the health component of EIA.

Alterations to the development proposals in order to reduce health risks may include selection of an alternative site of process changes intended to reduce health risks, reduction and control of emissions (e.g. filtering, abatement systems, etc.), or of emission of toxicity. Such measures are more likely to be accepted if they are proposed at an early stage, before the project design is finalised. This indicates the value of early involvement of the health impact team and cooperation with the developer. It is in the interests of the proponent to minimise negative impacts, to facilitate permitting procedures and to retain public goodwill.

An EIA may also consider actions in the community which would minimise health effects. An option would be to relocate any houses or establishments such as schools outside the area of risk. Less extreme actions might include provision of piped water supplies to residents at risk from contaminated well water, banning of consumption of fish from polluted rivers or reduction of air emissions from other sources to compensate for emissions from the new development.

5.8 CONCLUSIONS

This section has selectively reviewed examples of methods for EHIA. It makes no claim to be comprehensive. Other good EHIA methods may already be in existence, but have not yet come to general attention. The approach has been mainly to identify tools and information sources to help improve the health component of EIA and to illustrate some of the activities and methods of the EIA process for the benefit of health professionals. It is recognised that the future development of EHIA depends on cooperation and collaboration between the environmental and health professions.

The principal conclusion is that EHIA methods, as such, are not highly developed. Most of the methods in existence are simple methods for impact identification, such as checklists and matrices. The EHIA process will certainly benefit from the wider use and application of these methods. However, there is also a need for more complex methods for impact identification, such as networks, which do not so far appear to have been adapted for use in EHIA. These could facilitate the identification of indirect and long-term environmental health impacts. Another area of weakness is in prediction of exposure and effects on risk groups. It is frequently difficult to determine and measure the magnitude and distribution of environmental health impacts. More work is needed here, and this should be a major area of joint study by environmental and health professionals. Further specific guidance should also be developed on methods for interpreting significance and acceptability of environmental health impacts, and on available mitigating measures for different types of project.

REFERENCES AND BIBLIOGRAPHY

Bisset, R. (1980) Methods for environmental impact analysis: recent trends and future prospects, *J. Environ. Manage.*, **11**(1), 27–43.

Bisset, R. (1984) Methods for assessing direct impacts. In *Perspectives on Environmental Impact Assessment* (B.D. Clark et al. eds), D. Reidel, Dordrecht, pp. 195–212.

Canter, L. (1977) *Environmental Impact Assessment*, New York, McGraw-Hill.

Environmental Resources Ltd (1981) *Studies on Methodologies, Scoping and Guidelines, Volume 4, Methodologies*, Ministeri van Volksgezondheid en Milieu-hygiene and Ministerie van Cultuur, Recreatie on Maatschappelijk Werk, Netherlands.

Gilad, A. (1984) The health component of the environmental impact assessment process. In *Perspectives on Environmental Impact Assessment* (B.D. Clark et al. eds), D. Reidel, Dordrecht, pp. 93–103.

Giroult, E. (1984) *The Health Component of Environmental Impact Assessment*, paper presented at the International Seminar on Environmental Impact Assessment, University of Aberdeen, Scotland.

Kasper, R.G. (1980) Perspectives of risk and their effects on decision making. In *Societal Risk Assessment: How Safe is Safe Enough?* (R.C. Schwing and W.A. Albers Jr. eds), Plenum, New York, pp. 71–80.

Lowrance, W.W. (1980) The nature of risk. In *Societal Risk Assessment: How Safe is Safe Enough?*, (R.C. Schwing and W.A. Albers Jr. eds), Plenum, New York, pp. 5–14.

Parke, D. (1983) *Limits and Constraints to Scientifically Based Environmental Health Risk Assessment*, paper presented at WHO/EURO Working Group on Health and the Environment, Vienna.

PEEM Secretariat (1983) *Panel of Experts on Environmental Management for Vector Control (PEEM)*, Report of the Third Meeting, Rome, VBC/83.4, World Health Organization, Geneva.

Schneiderman, M.A. (1980) The uncertain risks we run: hazardous materials. In *Societal Risk Assessment: How Safe is Safe Enough?* (R.C. Schwing and W.A. Albers Jr. eds), Plenum, New York, pp. 19–37.

Slovic, P. et al. (1980) Facts and fears: understanding perceived risk. In *Societal Risk Assessment: How Safe is Safe Enough?* (R.C. Schwing and W.A. Albers Jr. eds), Plenum, New York, pp. 181–214.

United States Agency for International Development (1980) *Environmental Design Considerations for Rural Development Projects*, United States Agency for International Development, Washington, DC.

World Bank (1982) *The Environment, Public Health, and Human Ecology: Considerations for Economic Development*, World Bank, Washington, DC.

World Health Organization (various dates) *Environmental Health Criteria and Environmental Health Criteria Executive Summaries*, World Health Organization, Geneva.

World Health Organization (1982) *Rapid Assessment of Sources of Air, Water and*

Land Pollution, WHO Offset Publication No. 62, World Health Organization, Geneva.

World Health Organization (1983) *Selected Techniques for Environmental Management: Training Manual*, EFP/83.50, World Health Organization, Geneva.

World Health Organization Regional Office for Europe (1983) *Environmental Health Impact Assessment of Irrigated Agricultural Development Projects, Guidelines and Recommendations: Final Report*, World Health Organization Regional Office for Europe, Copenhagen.

Chapter 6

Data Bases and Information Sources

6.1 INTRODUCTION: ENVIRONMENTAL AND HEALTH DATA

Environmental health risks are managed by procedures based on the availability of a common set of data for chemical substances, physical factors such as noise and radiation, and other environmental hazards. On one level, the risks due to known exposure to many toxic substances are controlled via the setting of no (adverse) effect levels (air and water quality guidelines, occupational exposure limits, etc.), together with the establishment of appropriate monitoring schemes. As long as these administrative levels are not exceeded, it is assumed that there are no adverse health effects. The monitoring performed under such administrative mandates provides a rough description of the *environmental base-line*. As long as environmental base-lines of potentially toxic substances do not show upward trends it is assumed that environmental management is successful.

Except where human data are available, exposure guidelines and risk estimates are usually derived from experimental toxicity studies on laboratory animals, at high dose levels, which are then extrapolated by a range of different procedures to human environmental health risk factors. Due to the short-comings of this extrapolation, considerable effort has been placed in developing *environmental epidemiology*, as a tool to relate variation in incidence (patterns) with time or locality of some key diseases, with similar variation in the environmental base-line concentrations of priority chemicals. Studies of acute effects on aggregate populations can occasionally be made in the general environment (e.g. relationship between respiratory disease incidence in sensitive populations and episodes of high air pollution).

Some acute effects and many chronic effects can best be studied in terms of occupational exposures because of the potential for much higher absolute concentrations of the hazardous chemical, and the possibility of exposures which are better defined than those found in the outdoor environment. A drawback with this approach is that for certain effects (e.g. cancer induced by metals) there is a delay of 10–20 years between the first exposure and expression of the disease, making reconstruction of actual exposure situations extremely difficult.

The development and application of EHIA presupposes adequate, reliable, updated and accessible data covering the four major classifications of risk factors to health. They are:

(i)   *environmental exposures* to chemical substances and physical factors in the general, indoor and working environments and via air, food and water;
(ii) *lifestyle exposures* to risks from consumer products, drugs, pharmaceuticals, leisure activities, place, type and quality of residence, diet, use of alcohol, etc.;
(iii) *access to and quality of primary health care*; and
(iv) *individually determined biological susceptibility*.

Although such data are occasionally available via research data bases, there is no systematic collection of information which will permit a general assessment of the effects of environment on health within the region. The development of a collection of such data is, however, not sufficient, since it is also necessary to determine the relationships between individual and combined exposure to these risk factors and health outcomes.

The improvement and updating of such an *Environmental Health Data Base*, is an activity which should receive attention because an understanding of cause–effect relationships plays such an important role in EHIA activities. Since it can be expected that cause–effect relationships are similar throughout the region, such a unified data base would permit the developing and testing of hypothesis concerning tentative risk factors in various parts of the European Region. This would greatly facilitate the setting of priorities for environmental health risk management.

This chapter briefly introduces concepts, methods and sources of information which may be involved in the assessment of environmental health impacts within EIAs.

6.2 INDUSTRIAL DEVELOPMENTS

6.2.1 Environmental Health Factor Identification and Screening

Environmental health factors are those changes introduced by the proposed project, policy or plan which will affect the health and safety of the neighbouring population. In the case of industrial development projects, identification of environmental health factors will usually comprise determining sources of chemical and energy-related hazards. Over 60 000 chemicals are estimated to be in 'common use' and the number continues to rise, but complete toxicological data, including long-term experimental studies on carcinogenesis are available for less than 10 000 chemicals. If more rarely used chemicals are considered, a more pessimistic picture appears. As a consequence, a first problem in health impact assessment may be the lack of complete or adequate toxicological information.

6.2.2 Hazardous Substance Inventory

A fundamental requirement to allow identification of possible environmental health factors is an inventory of all substances which will be present at the site. The inventory should comprise chemicals:

(i)   used as inputs and raw materials in the planned industrial process, their amounts and storage procedures and siting;

(ii) produced in the process, their amounts and storage procedures and siting;

(iii) used in the process (for example, catalysts) and intermediate chemical products, their amounts, storage, procedures and siting;

(iv) contained as impurities in the above-mentioned classes;

(v)   which may be produced as a consequence of unplanned reactions, from the above-mentioned classes of compounds.

It is reasonable to expect to obtain parts (i), (ii) and (iii) from the project description, although some difficulties might arise from confidentiality of information in the case of proprietary industrial processes. Often these data may also be obtained by examining existing chemical plants analogous to the one under consideration.

In European Community (EC) member countries, this kind of information (at least in part) should be notified to competent public authorities as a consequence of the Directives on hazard classification of new chemicals (79/831/EEC and successive extensions and annexes) and of the 'Seveso' Directive (82/501/EEC). Data assembled to satisfy notification requirements of EC Directives and other similar national

Table 6.1
Classification of chemicals

Reactivity	Toxicity
Flammability	Acute toxicity
Explosivity	Subchronic toxicity
Corrosivity	Chronic toxicity
Radioactivity	Carcinogenicity
	Mutagenicity
	Teratogenicity
	Allergic response
	Infectiousness

regulations may be made available by national and international organi-sations in response to requests for assistance in information collection for a specific chemical project.

More difficulties may be expected in completing parts (iv) and (v) of the inventory. Detailed knowledge of materials and chemical production processes might be necessary in this case, as well as the knowledge of the substances whose possible chemical precursors are the chemicals listed in parts (i)–(iv). This kind of study will be necessary only in a limited number of cases, mainly when dealing with complex production processes involving particular classes of organic chemicals. In such cases, as in other aspects of health impact assessment, availability of specific expertise might be important.

6.2.3 Classification of Hazardous Properties of Chemicals
The second step is to obtain information on toxicological risks of each chemical in the inventory. In the base of chemicals in parts (i), (ii) and (iii), the collection of this information should not, in most cases, be difficult. Most chemicals used or produced in industrial processes (inputs and raw materials, intermediates and final products, etc.) are commercial products and, therefore, their characteristics should be known and should have been tested in some way, at least in part. The toxic and other hazardous properties, together with data useful to predict the environmental behaviour of such chemicals, should be more or less easily available in a large number of cases.

These parameters (or some of them) are used for the classification and labelling of chemicals on the market in most industrialised countries (e.g. EEC Directive 83/467/EEC) or as hazard indicators (Table 6.1).

The median lethal dose 50% (LD_{50}) is used to express quantitatively the toxicity of a substance (LD_{50} is the quantity of material that can be expected to kill 50% of the test animals under a prescribed set of conditions and with a given probability; it is advisable to make reference also to the confidence limits of such parameters). A detailed description of this parameter, as well as the methods used for its evaluation, is reported in a comprehensive WHO manual (WHO, 1986). LD_{50} values may be used for an easily comprehensible toxicity ranking of chemicals, as shown in Table 6.2. Classification systems based on the hazardous properties and toxicity rating of materials are widely used in most industrialised countries, and data may be available for more common substances.

6.2.4 Sources of Toxicological Data

The science of toxicology and environmental health hazard assessment is based on several scientific disciplines, requiring information from a range of different data and information sources. At first this may create some difficulties. As a rule, the library for health impact assessment should include manuals of general toxicology, occupational health, risk assessment procedures and environmental chemistry, among other sources depending on the type of development. A dictionary of common chemical compounds, reporting the basic chemical, physical, toxicological and hazardous properties is necessary, as is a guide to chemical name synonyms and international codes for chemicals (Chemical Abstract codes, codes adopted by the EEC). Such codes use numbers to identify chemicals precisely, avoiding possible errors due to different names in different languages and to differences in scientific, technical and common denominations. Moreover, these numbers facilitate accurate entry into computerised data banks.

The literature on the properties of chemicals is very large, and books and manuals have been published in many languages. As a first source it would be advisable for an EIA team to use manuals or chemical dictionaries published in their own language, to ensure better understanding of the information. International data services and foreign language books can complement locally available sources, providing possibly more detailed and authoritative data.

Data bases and data banks of interest for health impact assessment are all available in Europe and in most other industrialised countries. The different files are grouped under the names and classifications given by the producers. Several different data banks or bases may include among their components the same files, so that attentive analysis of contents is advisable to avoid repetition. Moreover, before investing in a computerised

Table 6.2
Toxicity rating

Probable human LD_{50} dose	Toxicity rating
5 mg/kg body wt	Super toxic
5–50 mg/kg body wt	Extremely toxic
50–500 mg/kg body wt	Very toxic
0·5–5 g/kg body wt	Moderately toxic
5–15 g/kg body wt	Slightly toxic
15 g/kg body wt	Practically non-toxic

data system, it should be tested appropriately, in order to ascertain if it is suitable.

6.2.5 Environmental Behaviour of Chemicals, Pathway Analysis, Environmental and Modelling

Environmental partitioning models are mathematically simple and are generally based on the solution of a system of equations (other types of models may be more complex). They make reference to the equilibrium concentrations of chemicals in air, water, water sediments, biota and soil organic carbon. The ratios between the various concentrations (soil/water, water/air, etc.) are defined by partition constants, which may be obtained from the chemical/physical data. Small computers may be used for the calculations required by the models. The programmes may be easily written from the mathematical relationships reported in the papers describing the models or may be requested from the authors.

Environmental concentrations estimated in this way may be used as input for the further analysis, which requires more specific models and procedures. For instance, if a chemical is expected to affect mainly environmental water (for example, because it is highly water-soluble), it is important to predict its mobility in water. Supposing that the hypothetical contamination is initially in the upper soil layer, water contained in the soil and surface water close to the contaminated area will soon be affected. The next step is to predict the movement and uses of surface water (possibility of contamination propagation, effect on ground water, absorption of water by plants and their possible contamination, humans and animals that may drink polluted water, etc.). This simple example shows that the analysis progressively moves from theoretical modelling and intrinsic properties of chemicals to a study gradually more dependent on the real characteristics of the particular environment. It is difficult to give general criteria for this latter type of analysis, which requires knowledge of the system and of the possible pathways followed by the contaminant before reaching man. It is advisable to study carefully, indications emerging from past experience and from research results in cases not necessarily identical to the one under study. For instance, it may be expected that two chemicals, with similar physical/chemical parameters, will take analogous pathways in the same environment. If experimental data are available that describe the behaviour of one of them, an extrapolation will be possible.

6.2.6 Preliminary Health Hazard Matrix
Information collected and analysed in the phase of environmental health factor identification may be summarised in an appropriate matrix, a 'preliminary health hazard matrix'. For each hazardous compound, the LD_{50} values or alternative acute toxicity rating values and the chronic toxicity data (or in their absence, at least the subchronic toxicity data) should be reported together with mutagenicity data and, whenever possible, with the carcinogenicity and teratogenicity potentials, at least in qualitative terms (for example, carcinogen/non-carcinogen) and possibly in quantitative terms (as derived from dose-response relationship and low-dose extrapolation procedures). The reactive properties and other relevant characteristics of materials and substances should also be entered in the matrix, together with indications on the environmental compartments which are expected to be mainly affected after environmental emission (based on physical/chemical characteristics). It is possible that all the information for each chemical is neither available nor can it be estimated from other available data. In these cases the matrix will serve the valuable purpose of highlighting areas of uncertainty.

6.2.7 Routine and Accidental Release of Toxic Materials
Any project, however toxic the substances present within the plant, would have no community health effects if the materials were not released to the environment. Environmental health impacts arise only when the surrounding population comes into contact with toxic or hazardous substances. Therefore, health impact assessment is concerned with the hazardous nature of materials and their behaviour in the environment, only in so far as they might be released. Description of the quantity and nature of air emissions, effluent discharges and solid waste is part of the normal EIA process, as is the prediction of the dispersion and concentration of waste materials in the environment. The release of materials

can be divided into those which are routine and can be anticipated and those which will occur only as a result of accident or some other failure.

6.2.8 Dispersion of Released Materials

Having estimated the quantities of materials released into the environment, the next stage is to determine how they will be dispersed and what concentrations might occur. The practice of air and water dispersion modelling is well developed and is a standard component of EIA. The outputs of these models, relating to chemicals identified as environmental health factors, will be important in assessing health effects. Dispersion model results are usually presented in terms of maximum concentrations under adverse weather or water conditions. In the case of air emissions, the distance to the point of maximum ground level concentration is also given. The models become more complex when attempting to predict the behaviour of emissions from more than one source. When the models are to be used for health impact assessment, they may be asked to provide more than one maximum figure; for example, ground level concentrations at sensitive points, such as housing areas and hospitals, would be significant.

A particular problem when assessing the impacts of the release of a number of substances, both on the environment and on human health, is the lack of data on the combined effects of chemicals. The effect of two or more substances together may be additive, synergistic or antagonistic, and knowledge of the combined effects would greatly assist interpreting the results of air and water quality modelling.

6.3 GUIDANCE MANUALS

6.3.1 US AID Manual

The United States Agency for International Development (US AID) (1980) produced an EIA manual entitled *Environmental Design Consideration for Rural Development Projects*. This is one of the few EIA manuals which includes a comprehensive range of health considerations. The main method used is the questionnaire checklist, answers to specific questions relating to the particular project being assessed, and once an initial question has been answered in the affirmative, additional questions investigate the nature of particular impacts in detail. The actions needed to provide the type of information required ensure a reasonably thorough consideration of project characteristics and environmental health factors, so some impacts may be missed.

Major Tropical Diseases \ Rural Development Projects	Rural Roads	Electrification	Water Supply/Sanitation	Small-Scale Irrigation	Small-Scale Industry
1. African Sleeping Sickness	o	o		o	
2. Dysentery (Bacillary & Amoebic)			o	o	
3. Chagas' Disease	o	o			o
4. Cholera			o		o
5. Dengue	o	o		o	
6. Filariasis [a/]	o	o	o	o	o
7. Guinea Worm Disease			o	o	
8. Hemorrhagic Fever					o
9. Hookworm Disease [a/]	o	o		o	o
10. Malaria [a/]	o	o	o	o	o
11. Leishmaniasis [a/]	o	o			
12. Leptospirosis			o	o	o
13. Onchocerciasis	o	o		o	
14. Plague	o	o			
15. Rabies	o	o			o
16. Relapsing Fever [a/]	o	o		o	
17. Schistosomiasis			o	o	o
18. Typhoid & Paratyphoid Fevers	o	o			o
19. Shrub Typhus	o	o		o	
20. Yellow Fever	o	o			

a/ One of six major diseases in the World Health Organisation
Special Program for Research and Training in Tropical Diseases

Fig. 6.1. Tropical diseases likely to be affected by rural development projects.

The manual also contains a useful appendix indicating which tropical diseases are likely to be affected by which types of rural development project (Fig. 6.1). The matrix method, which is essentially a combination of two simple checklists, facilitates systematic identification of impacts. In this case the vertical axis is a checklist of the major tropical diseases, further details of which are given in the text. The horizontal axis is a checklist of rural development projects. By relating the checklists to one another, the likely disease effects of the projects have been systematically identified. The method is somewhat unusual in that it provides information as well as structuring it. It would be useful in very early stages of project planning for rural development projects in tropical countries.

Effects on inhabitants of project area

Communicable disease
Housing and sanitary facilities
Dietary change
Effects on groundwater
Changes in ecological balance
Changes in agriculture
Increased risk of road accidents
Risks to community health from certain industrial processes

Effects on workers

Work accidents
Exposure to chemical and physical hazards
Exposure to local diseases
Nutritional status of workers

Indirect effects

Introduction of new disease vectors
New infection or reinfection of existing vectors
Increased propagation and spread of existing vectors

Effects on existing health services

Fig. 6.2. Checklist of environmental health factors that may be affected by development projects. Based on World Bank (1982).

Obviously neither of these particular methods is suitable for universal application, but they do illustrate two available mechanisms for impact identification. Moreover, both questionnaire checklists and matrices are very adaptable tools for impact identification, in that different ones can be designed for different situations. Other examples will be given later.

6.3.2 World Bank Handbook
The World Bank (1982) publication, *The Environment, Public Health and Human Ecology* is 'designed to provide guidance in the identification, detection, measurement and control of environmental effects'. It includes a number of EHIA methods.

First, it provides what is in effect a general checklist of environmental health factors that may be affected by development projects (see Fig. 6.2). Simple checklists of this sort help ensure that no important impacts are omitted from an assessment, but have few other advantages. They provide less information than questionnaire checklists, and are less thorough than matrices.

In addition, the handbook includes a 'Checklist of Environmental Considerations'. Like the US AID checklist, this is a questionnaire checklist. It is actually not one checklist, but several, for it is subdivided by project type. The following project types are included:

(i)   energy resource systems: oil and natural gas, coal, uranium
(ii) industrial development
(iii) transportation: airports, ports and harbours
(iv) sewerage and sewage treatment

For each project type, questions are asked under a number of different headings, including environmental/resource linkages, project design and construction, site assimilative capacity, waste management, operations, socio-economic factors, health impacts and long-term considerations. This method has a number of advantages. Its project-specific approach is good, for it facilitates identification and encourages investigation of project-specific impacts. The examination of different project actions and phases is also good, as is the explicit consideration of health impacts.

Again, the weakness is in the failure to systematically relate project actions to environmental factors. For example, only general questions concerning health impacts are asked, rather than questions related to design, construction and operational phases in turn.

6.3.3 WHO Environmental Health Criteria

The WHO Environmental Health Criteria (WHO, various dates) can be used as an EHIA method or simply a source material. The distinction is unimportant. The criteria can assist execution of a number of activities in the EHIA process, including identification of health risk groups and prediction of how they will be affected by exposure to environmental health factors, definition of significance and acceptability of adverse health impacts, and identification of mitigation measures.

The criteria represent the output of the WHO Environmental Health Criteria Programme, which is part of the more comprehensive International Programme on Chemical Safety (IPCS). The aim of the WHO Environmental Health Criteria Programme is to assess existing information on the relationship between exposure to environmental pollutants (or other physical and chemical factors) and man's health, and to provide guidelines for setting exposure limits consistent with the protection of public health. The environmental pollutants included in the series have been carefully selected according to the following priorities:

(i)   severity and frequency of observed or suspected adverse effects on human health

1. Mercury
2. Polychlorinated Biphenyls and Terphenyls
3. Lead
4. Oxides of Nitrogen
5. Nitrates, Nitrites, and N-Nitroso Compounds
6. Principles and Methods for Evaluating the Toxicity of Chemicals, Part 1
7. Photochemical Oxidants
8. Sulphur Oxides and Suspended Particulate Matter
9. DDT and its Derivatives
10. Carbon Disulfide
11. Mycotoxins
12. Noise
13. Carbon Monoxide
14. Ultraviolet Radiation
15. Tin and Organotin Compounds
16. Radiofrequency and Microwaves
17. Manganese
18. Arsenic
19. Hydrogen Sulfide
20. Selected Petroleum Products
21. Chlorine and Hydrogen Chloride
22. Ultrasound
23. Lasers and Optical Radiation
24. Titanium
25. Selected Radionuclides
26. Styrene
27. Guidelines on Studies in Environmental Epidemiology
28. Acrylonitrile
29. 2,4-Dichlorophenoxyacetic Acid (2,4-D)
30. Principles for Evaluating Health Risks to Progeny Associated with Exposure to Chemicals during Pregnancy
31. Tetrachloroethylene
32. Methylene Chloride
33. Epichlorohydrin
34. Chlordane
35. Extremely Low Frequency (ELF) Fields
36. Fluorine and fluorides
37. Aquatic (Marine and Freshwater) Biotoxins
38. Heptachlor
39. Paraquat and Diquat
40. Endosulfan
41. Quintozene
42. Tecnazene
43. Chlordecone
44. Mirex
45. Camphechlor
46. Guidelines for the Study of Genetic Effects in Human Populations
47. Summary Report on the Evaluation of Short-Term Tests for Carcinogens (Collaborative Study on In Vitro Tests)
48. Dimethyl Sulfate
49. Acrylamide
50. Trichloroethylene
51. Guide to Short-Term Tests for Detecting Mutagenic and Carcinogenic Chemicals
52. Toluene
53. Asbestos and Other Natural Mineral Fibres
54. Ammonia
55. Ethylene Oxide
56. Propylene Oxide
57. Principles of Toxicokinetic Studies
58. Selenium
59. Principles for Evaluating Health Risks from Chemicals During Infancy and Early Childhood: The Need for a Special Approach
60. Principles for the Assessment of Neurobehavioural Toxicology
61. Chromium (in preparation)
62. 1,2-Dichloroethane
63. Organophosphorus Insecticides – A General Introduction
64. Carbamate Pesticides – A General Introduction
65. Butanols – Four Isomers
66. Kelevan
67. Tetradifon
68. Hydrazine
69. Magnetic Fields
70. Principles for the Safety Assessment of Food Additives and Contaminants in Food
71. Pentachlorophenol
72. Principles of Studies on Diseases of Suspected Chemical Etiology and Their Prevention
73. Phosphine and Selected Metal Phosphides
74. Diaminotoluenes
75. Toluene Diisocyanates
76. Thiocarbamate Pesticides – A General Introduction
77. Man-made Mineral Fibres

Fig. 6.3. Titles in the series of WHO Environmental Health Criteria.

Fig. 6.3.—*contd.*

78. Dithiocarbamate Pesticides, Ethylenethiourea, and Propylenethiourea – A General Introduction
79. Dichlorvos
80. Pyrrolizidine Alkaloids
81. Vanadium
82. Cypermethrin
83. DDT and its Derivatives – Environmental Aspects
84. 2,4-Dichlorophenoxyacetic Acid – Environmental Aspects
85. Lead – Environmental Aspects

86. Mercury – Environmental Aspects
87. Allethrins
88. Polychlorinated Dibenzo-*para*-dioxins and Dibenzofurans
89. Formaldehyde
90. Dimethoate
91. Aldrin and Dieldrin
92. Resmethrins
93. Chlorophenols
94. Permethrin
95. Fenvalerate
96. d-Phenothrin

(ii) ubiquity and abundance of the agent in the human environment

(iii) persistence in the environment and possible environmental transformations or metabolic alterations important from the point of view of health risk

(iv) size of population exposed and/or selective exposures of critical groups of population

The titles in the series of WHO Environmental Health Criteria are listed in Fig. 6.3. For each title there is a full criteria document, comprising a detailed scientific review of sources and exposure levels, sensitive populations, available evidence on health effects including dose-incidence and dose-response relationships, and guidelines on exposure limits. For each title there also exists an executive summary highlighting the information contained in the document for those who need to know the health issues, but not the scientific details. The executive summaries are intended to facilitate the application of guidelines on exposure limits in national environmental protection programmes.

The criteria contain the best available technical summaries of health effects of environmental pollutants, and are easily accessible. One of the potential roles of EHIA is to ensure that the information and guidelines provided by the criteria are properly integrated and applied in the planning and assessment of new development projects.

REFERENCES AND BIBLIOGRAPHY

Birley, M.H. (1991) *Forecasting Potential Vector Borne Disease Problems of Irrigation Schemes*, World Health Organization, Copenhagen.

Martin, J. (1986) *The Health Component of EIA*, presented at International Seminar on EIA, University of Aberdeen, Scotland.

United States Agency for International Development (1980) *Environmental Design Considerations for Rural Development Projects*, Washington, DC.

World Health Organization (1978) *Principles and Methods for Evaluating the Toxicity of Chemicals, Part* 1, (Environmental Health Criteria 6), WHO, Geneva.

World Health Organization (various dates, 1976 onwards) *Environmental Health Criteria Series*, WHO, Geneva.

World Health Organization (1987) *Health and Safety Component of Environmental Impact Assessment*, Environmental Health Series No 15, WHO, Copenhagen.

Chapter 7

Research Needs: Future Development

7.1 WHY AND HOW TO STRENGTHEN HUMAN HEALTH CONSIDERATIONS IN ENVIRONMENTAL IMPACT ASSESSMENT

There is general agreement on the importance of three main categories of environmental impact which are reflected in the objectives of the European Economic Community's Programme of Action for the protection of the environment. They are:

(i)   protecting human health;
(ii) safeguarding the resources on which life depends; and
(iii) protecting the natural environment.

Until now, most impacts on human health have not been directly assessed. For example, practice in the United States was to assess only the impact on the environmental quality parameters, then if the resulting value of these parameters was not above the legally established environmental standard, it was considered that no significant human health impact would result. As the environmental assessment process has evolved in the United States, coupled with a greater concern about toxic chemicals and hazardous wastes, the US Environmental Protection Agency (EPA) has increasingly been dealing with many chemical substances for which there may be no established health tolerances, and which can only be dealt with on a risk assessment and management basis. As a result, there is currently a greater use of health risk assessments in EIA. The World Health Organization also considers that within the EIA process there is a need to assess as closely as possible feasible human health impacts.

7.2 WHERE EIA STANDS IN THE SCOPE OF ENVIRONMENTAL HEALTH ACTIVITIES

There are several views on the definition of environmental health and of its scope. However, a consolidated diagram (Fig. 7.1) prepared by Dr Zoetemann, Director of the Environmental Health Institute in Bilthoven, The Netherlands, shows environmental impact assessment as the interface between natural resources and the production of activities on one side, and the main component of the human environment such as air, water and food on the other. Radiation, vibrations and animal vectors of diseases are added to the latter. Health risk assessment stands at another interface between the main component of the environmental pathways of exposure and the human being itself.

7.3 WHY THE NEED TO STRENGTHEN HEALTH CONSIDERATIONS IN EIA

The general accepted approach to environmental impact assessment has been as follows:

The first step
To assess the impact of the project on primary environmental parameters such as the concentration of a specific chemical in natural waters resulting from the project operation.

The second step
To assess secondary or tertiary impacts such as the biological concentration of the above specific chemicals in a particular aquatic organism.

The third step
To compare the value of environmental parameters resulting from primary, secondary or tertiary impacts, to the legal or recognised environmental quality standards, similar to the action taken by the EPA referred to in Section 7.1.
This approach is considered to be insufficient for the following reasons:

- Available environmental standards deal mostly with indicator parameters easy to measure and to standardise, and less commonly with the more dangerous compounds. Air quality standards, for example,

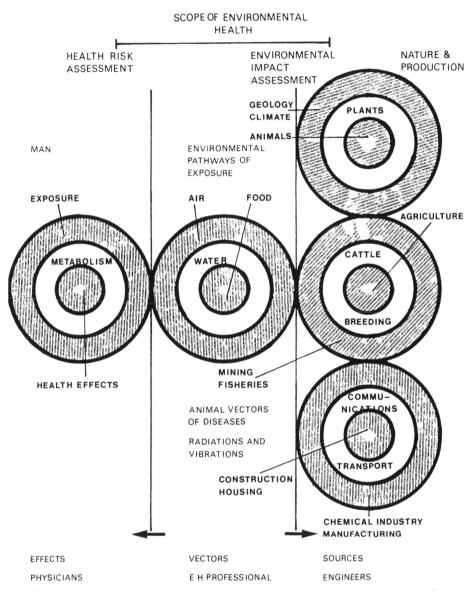

Fig. 7.1. EIA as the interface between national resources and the production of activities and the main components of the human environment.

First Priority

Possible damage on human health and/or safety

Second Priority

Possible loss of economically valuable natural resources such as
fresh water, agricultural soil, forests, fish, building materials etc

Third Priority

Possible loss and/or damage on ecological or cultural values such as
endangered species, landscapes, archeological sites, historical
monuments, etc.

Fig. 7.2. Classification of environmental health impacts.

will refer to dust particulates, smoke, SO_2, or NO_2 which clearly have
a health impact but are selected because they are the most common
and ubiquitous of substances, and because of their availability to
monitoring. Chemicals recognised for being carcinogenic such as
certain halogenated polycyclic aromatic compounds included in
vehicle exhaust gases are not generally monitored due to their low
concentration. Consequently there is often no air quality standards
dealing with aromatic compounds. The same applies to a significant
part of toxic chemicals leaching in underground waters from
hazardous waste disposal sites.

● There is a minimum understanding on the part of the general public
concerning the real significance of environmental health standards
(e.g. public health experts and toxicologists know or should know
this). For example, when a standard is approved for the limit value of
a specific substance in drinking water, the tendency is to believe that
below this standard there is no impact at all on human health, and
that above this standard the worst health damages may arise. In fact,
a standard is often only a compromise resulting from the risk
management process, a compromise between maximum efforts to
protect human health and what is feasible technically and economically in terms of water purification. For example, the following
health criteria have been recommended for the design of Drinking
Water Quality Standards related to carcinogenic substances.

The recommended concentration for each specific chemical is computed
according to the results of scientific data available on its biological impact,
and on the average assessed human exposure to the chemical, so that
exposure to this specific chemical will not result in an additional risk of
cancer higher than 1/100 000 (estimated through conservative risk as-

sessment procedures). The *assessment of human exposure* to environmental parameters, the value of which may be modified by development projects, is what is missing in the general approach to environmental and health impact assessment.

7.4 HOW TO STRENGTHEN HEALTH CONSIDERATIONS IN EIA

From the public health viewpoint, there is a need to consider the following priority categories of environmental impact, and the possible damage or loss shown in Fig. 7.2.

First priority
Impacts which may affect human health and/or safety.

Second priority
Impacts which may damage economically valuable natural resources (including water, soil, forest, fishes, building materials, etc.).

Third priority
Impacts which may damage other ecological or cultural values (endangered species, archaeological sites, historical monuments, beautiful landscapes).

To strengthen health considerations in the EIA process, further steps are required in addition to the first three steps described in Section 7.3, together with a modification to Step 3. The modification should include the screening between impacts which may affect human health and/or safety and other environmental impacts, based upon well established epidemiological or toxicological facts, and consolidated in the format of Environmental Health Criteria Documents.

The fourth step would provide the key to human health impacts through the assessment of human exposures to environmental factors or parameters of public health significance, the value of which is modified by the development project. The question of human exposure assessment location (HEAL) is extremely important and is discussed later.

The fifth step entails the assessment of how far the project will change the exposed population, the percentage of vulnerable persons, and especially of highly vulnerable persons or risk groups, in relation to the specific biological process involved. If the vulnerable group consists of aged persons, it may happen that certain development projects involving resettlement of populations will increase the percentage of aged persons among the exposed population, because old people are reluctant to move. The question of different sensitivity of human beings to the same exposure to environmental health factors should not be confused with other problems which is that some environmental health factors may increase the sensitivity of human beings to infectious agents. For example, the irritating effects of particulates on respiratory air tracts may increase the sensibility of bronchi to gaseous air pollutants.

The sixth step entails the assessment of how far the project will offset human health in terms of mortality and morbidity. To achieve this sixth step, there is a need for adequate knowledge of dose–response relationships. Unfortunately, full knowledge of these dose–response relationships in all cases of chemical exposure are not yet available. Reference to *risk assessment* in this context covers scientific research based partly on human clinical studies, partly on epidemiological studies, but mostly on laboratory animal experiments. Continued research progress in this field is intended to build up dose–response curves. The main limitation, however, is how far the results of animal experiments may be transferred to humans and also how far the impacts observed at high doses may be extrapolated to low doses. The six main steps are shown in Fig. 7.3.

Actual accepted knowledge are consolidated in the format of WHO Guidelines for Drinking Water Quality dealing with the health effects of several important solvents or pesticides, and in the format of WHO European Guidelines of Air Quality dealing with twenty-three indoor and outdoor air pollutants.

As EIA is a practical process it is not possible to undertake additional scientific research. Consequently, conclusions should be based on presently accepted scientific conclusions. The sixth step, therefore, may be considered as belonging to another process than EIA, which is the health risk assessment process indicated in Fig. 7.1 prepared by Dr Zoeteman. A seventh step is required to evaluate the result of the health impact assessment. This seventh step is also called health risk management process which is schematically described in Fig. 7.4.

1	To assess direct impacts on environmental parameters
2	To assess indirect impacts on environmental parameters
3	To screen environmental parameters which have a health significance (EH factors)
4	To assess increase of exposure
5	To assess increase in risk-group populations
6	To assess health impacts (mortality and morbidity) (Link to health risk assessment followed by health risk management)

Fig. 7.3. Environmental health impact assessment—main steps.

7.5 HUMAN EXPOSURE ASSESSMENT, EXPOSURE PATHWAYS, RISK GROUPS

As indicated above, the specific aspects of environmental and health impact assessment are the additional steps that need to be taken to assess human exposure with due regard to the high sensitivity of risk groups.

Despite the fact that today's priority concern is related to toxic chemicals, it is important to remember that environmental factors of health significance may be physical such as radiation, noise or vibration, or biological such as animal vector if infectious or parasitic diseases, or sensory and psychological such as odours, visual offences, loss of amenity and recreational facilities.

It is relatively easy to assess the number of people exposed for example, to excessive noise levels and perhaps to screen among them the sensitive groups. It is also fairly easy to assess the number of people living inside the flying range of an insect vector of infectious or parasitic disease. It is more complicated, however, to evaluate human exposure to chemicals because several pathways of exposure should be considered. The main pathways are:

- the direct contact (between the skin and soil, dust, liquids, radiation);
- inhalation (the air and the lungs);
- ingestion (food or water and the digestive tract);
- exceptionally — inoculations (through the skin, the flesh and the veins).

The screening between these different pathways to identify the most significant one can be done with due consideration to the physical properties of the specific chemical. For example, a chemical of high volatility will be found in the air, a chemical of high water solubility will be found in water, etc. To assess the number of people exposed to, for

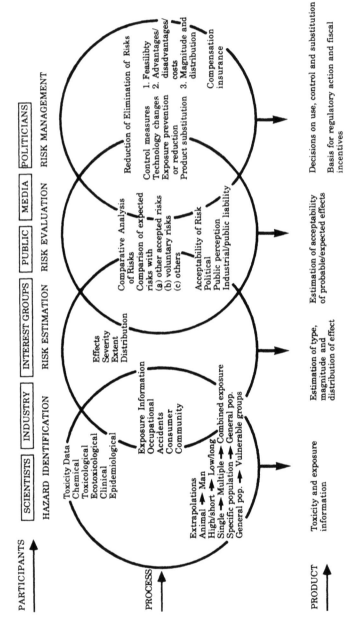

Fig. 7.4. Chemical risk management.

example, chemical leaching out of hazardous waste disposal sites, there is a need to consider the possibility of a combination of these chemicals in the environment and the range covered by the chemical according to its principal pathway, also called route of exposure. For example, a chemical of high water solubility will leach in surface or groundwater. In surface water there may be a possibility of chemical reaction with other compounds or of biological concentration through the biological food chain.

If the chemical leaches in groundwater, the speed of its spreading in the aquifer as well as its spatial concentrations may be computed on the basis of hydrogeological data with an adequate mathematical model. Then the population exposed will be those served by drinking water extracted from this aquifer at points of extraction where the computed concentration appears noticeable. This example shows that the assessment of human exposure to chemicals is a complicated but feasible process.

The assessment of human exposure should be completed by the identification of risk groups inside the exposed populations. To make it simpler, it is generally considered that the identification of biological or genetic risk groups is sufficient. This identification will later help the design of mitigation measures which are discussed below before returning to the question of risk assessment.

7.6 DOSE–RESPONSE RELATIONSHIP OR RISK ASSESSMENT FOR CHEMICAL OR PHYSICAL FACTORS

Established epidemiological knowledge provides accurate data on the dose–response relationship for biological pathogens, efficient infectious doses for the main pathogens parasites, bacteria or viruses.

As far as chemical or physical factors are involved, it has been established in some cases that the significant dose is the immediate concentration, and in other cases that it is the cumulative dose. Dose–response relationships are reasonably well known for high doses of physical factors such as noise or radiation but are less certain in relation to prolonged exposure to low dose levels.

In relation to chemicals, proven data from toxicological research upon animals are for example, the 50% lethal dose (LD_{50}) for different pathways such as skin contact, ingestion or inhalation. The results change with the animal species tested, which opens a wide range of discussion on the validity of extrapolation of the results to human species.

Other sources of information are epidemiological results when applicable, or occupational health clinical results. However, it is important to remember that occupational health most often deals with relatively high doses applied to non-sensitive individuals because risk groups are prevented from going into hazardous occupations. This is known as the healthy worker effect.

For the many chemicals, the dose–response relationship for short-term health impacts and relatively high doses is well known. However, as new chemicals are developed each year, this optimistic statement only applies to a minority of chemicals.

In many cases, the long-term health effects of exposure to low dose of chemicals are not well known. These long-term effects may be carcinogenic, teratogenic or genetic. The International Agency for Cancer Research is continuing to extend today's knowledge related to the carcinogenicity of selected chemicals and a lot of results are already available.

7.7 EVALUATION OF ENVIRONMENTAL HEALTH IMPACT ASSESSMENT AND MITIGATION MEASURES

Environmental and health impact assessment exercises are not theoretical, they are practical and should lead to decision-making.

When a potential health hazard has been identified through the screening process referred to in Step 3, it should be quantified through the scoping process of Steps 4 and 6. Then the results of the screening and scoping, which means the quantifiable health impacts, should be subjected to evaluation through Step 7, which may be called *definition of acceptable risks or risk management*. While it is not possible deal in-depth with the risk management theory, either all quantified health impacts fall within the range of acceptable risks and the project is acceptable to the public and health authorities, or it is not. If no better evaluation tools are available, it is reasonable to consider that if the value of environmental factors of health significance modified by the development project, stay compatible with the accepted environmental health quality standards, it means that the health impact is inside the range of acceptable risk. This conclusion is compatible with the Environmental Protection Agency practice referred to in Section 7.1.

According to the risk management criteria used by public authorities, if one or several quantified health impacts are clearly out of the range of acceptable risks, the development project cannot be approved unless satisfactory mitigation measures or alternative project designs are proposed.

The first category of mitigation measures include all those which are used to decrease the severity of offensive environmental impact, for example the following:

- change the site of the project;
- change its design or its proposed technology;
- add pollution abatement systems;
- provide measures to control impacted environmental factors and/or parameters and so on.

A second category of mitigation measures is specific to the environmental impact of high significance for human health. They include:

(i) all necessary measures to decrease the number of population exposed and the intensity of exposure;
(ii) the screening of sensitive persons and special protection measures for risk groups;
(iii) compensatory public health measures.

By way of examples, building a water reservoir under dry and warm climates will increase the mosquito population and if insects or humans are infected, they may spread mosquito-borne diseases such as malaria or yellow fever. Mitigation measures of the first category may be chemical or biological control of mosquitoes. Mitigation measures to decrease population exposure may be resettlement of rural population far from the project site. Compensatory public health measures may be chemoprophylaxis for malaria or mass-immunisation for yellow fever.

Compensatory public health measures related to health risks associated with the spreading of chemicals in the environment will be designed in accordance with the exposure pathway related to each chemical. If the pathway is underground water, the compensatory measures may be either the protection of private shallow wells or the active carbon treatment of water extracted from public wells. If the pathway is food, the compensatory public health measure may be the prohibition of sensitive crops in the exposed parameters, etc.

The design of satisfactory *mitigation measures* will be the eighth step of the environmental and health impact assessment process. The final or ninth step, will be decision-making. If the public health authorities are convinced that the proposed mitigation measures will be efficient enough to reduce the magnitude of quantified health impacts inside the range of acceptable risks, they will accept the project (see Fig. 7.4).

To achieve proper consideration of health impacts in EIA, it is necessary to obtain full collaboration from public health professionals. It is, therefore, necessary to convince public health experts that they should contribute to environmental impact assessment exercises of all development projects susceptible to having an influence on human health and environmental health. If EIA is left to environmentalists alone, public health professionals cannot reasonably expect that possible impacts on human health and safety will be adequately assessed.

The EIA process is especially adequate for dealing with land use planning and its environmental and public health consequences. It was originally developed as a decision-making tool for siting economic development projects such as hazardous industrial plants, but it may also be used to assess the environmental and health consequences of economic development policies such as energy production, intensive agriculture, transportation, urban development, etc.

The involvement of public health services in EIA is an intersectoral process and was strongly recommended at the World Health Assembly Session in 1986, through Resolution WHA/37.22 — Intersectoral Co-operation in National Strategies for Health, following the 1982 Resolution WHA/35.17 on Health Impact of Development.

To be able to contribute positively to EIA, public health professionals should be trained in EIA concepts and methods. This is why the WHO Regional office for Europe annually sponsors training courses organised by the Centre for Environmental Management and Planning at Aberdeen University.

7.8 WORLD HEALTH ORGANIZATION ACTIVITIES IN ENVIRONMENTAL IMPACT ASSESSMENT

The potential of EIA to promote good health in communities affected by development policies and projects is actively encouraged by the support given by WHO for EIA training and research. For example, to overcome the perceived weakness of the health component of EIA, WHO/EURO developed a separate procedure for environmental health impact analysis (EHIA) which focuses on environmental impacts which have a significance for health. Attempts have been made to apply the EHIA procedure to various types of development (Fig. 7.5).

Step 1	Project Description	Toxic substances inventory
Step 2	Assessment of primary impacts on environmental parameters	Normal EIA process
Step 3	Assessment of secondary and tertiary impacts on environmental parameters	Normal EIA process
Step 4	Identification of impacted environmental parameters with health effects (EHF). Preliminary assessment of environmental health factors	Epidemiological, toxicological information
Step 5	Prediction of exposure to environmental health factors	Environmental behaviour, pathway analysis
Step 6	Identification of health risk groups	Population analysis, lifestyle, diet
Step 7	Estimation of predicted health impacts	Chronic effects of routine releases, hazard analysis of accidental events
Step 8	Identification of mitigation measures to prevent or reduce significant adverse health impacts	Process changes and/or protection measures for population
Step 9	Final decision on acceptability of adverse health impacts and whether or not the project should proceed	Normal decision-making process, with due weight given to health effects

Fig. 7.5. The EHIA process applied to chemical manufacturing facilities.

7.8.1 WHO Working Group on the Health and Safety Component of Environmental Impact Assessment

In 1986, WHO took the initiative in inviting a number of experts to a Working Group Meeting in Copenhagen to discuss the health and safety component of EIA. The Working Group recognised the importance of the EHIA procedure in drawing attention to environmental health effects of policies and projects, but considered that EHIA should always be an integral part of EIA and that assessment of health effects could be best improved within the context of EIA as a whole.

Separate assessment procedures for health impacts may imply that they are in some way different from other impacts, and they may consequently fail to receive appropriate attention both during assessment and in the final decision. Health impact assessment should be integrated within EIA and direct and indirect effects on environmental health should be identified, predicted and evaluated alongside other impacts.

The health and safety component of EIA will be strengthened by ensuring that those who conduct EIAs are always aware of the possibility that there may be health effects and that techniques for assessing those effects and sources of relevant expertise and data become more widely known.

7.8.2 The Working Group Meeting Identified a Number of Ways in which the Health Component of EIAs could be Strengthened

Among the considerations were:

- specifications for the content of EIAs and general guidance material should include the requirement to assess environmental health impacts where applicable;
- individual organisations with environmental health expertise should be included in 'scoping' activities to ensure that health and safety effects are considered from the outset. Where there are no formal 'scoping' procedures, advice from appropriate experts should be sought at the beginning of the EIA process;
- the purpose of the EIA should be made clear; approaches to impact assessment and presentation of results will vary between documents prepared for the general public and technical material for the decision-makers;
- where 'scoping' indicates that significant health issues may arise, health professionals should be included in the assessment team;
- there should be close cooperation with the proponent to ensure all relevant information is provided and to allow the development proposals to be modified in the light of health factors;
- data inputs for health impact assessment for a project should include full materials inventory, nature of materials, environmental behaviours and pathway analysis, population information including number, characteristics, location and lifestyle, and hazard assessment;
- public health authorities should be involved at all stages of environmental health impact identification and assessment;
- the assessment of alternatives is crucial, albeit that a number of options would have been considered at the pre-feasibility planning stage by the proponent. However, alternatives must be reasonable and, in the case of a project EIA, concern location, design, no development options rather than alternative policy strategies;
- in most countries, EIAs are public documents and the health and safety component should also be public. When health impact assessments are treated as confidential, the information may never-

theless become public knowledge and be misused. Careful presentation of health impact information ought not to create unjustified alarm, but at the same time EIAs should not become public relation exercises for the development proponent. There is a need to increase knowledge and experience in communicating health effects to the public;

● community and public participation in the EIA process; particularly regarding the health component, would be facilitated by knowledgeable mass media;

● whenever possible environmental health effects should be quantified;

● air and water quality standards are useful indicators of acceptable concentrations of pollutants but they should not be used without explanation and interpretation.

7.8.3 The Working Group Agreed Four Principles to Guide the Assessment of Health Effects within EIA

(1) One of the fundamental considerations in approval of development projects, policies and plans should be the health of communities affected by them.

Current experience suggests that health has not received sufficient attention in such decisions and this has led to real costs being incurred in both social and economic terms. Consequently, the second principle is that:

(2) Greater consideration should be given to the consequences of projects, policies and plans for human health.

EIA is used to provide information for the decision-makers on the consequences of action. The evidence suggests that in the past EIAs have not generally provided adequate information on health impacts. The third principle is therefore that:

(3) EIA should provide the best available factual information on the consequences for health of projects, policies and plans.

It is recognised that EIA is ideally a procedure open to the public and the fourth principle is that:

(4) Information on health impact should be available to the public.

7.8.4 To promote the above Principles, Three Objectives were Proposed

The first objective is:

● *to increase awareness of the potential benefits of improving consideration of health effects in EIA*

(1) To inform health professionals, that is all scientific and medical personnel with relevant experience, of preventative opportunities offered by EIA through:
— dissemination of information in professional journals, newsletters and training materials
— training activities, conferences, seminars and courses
— sponsoring participation of relevant health professionals in EIA teams (a) within recognised training courses, and (b) practising doctors as in-service training.

(2) To persuade proponents and decision-makers of the dangers of not considering the health impacts of their decisions through:
— dissemination of information, by official letter, on the consequences of failing to consider health effects of projects, policies and plans implemented in the past
— encouraging the inclusion of health issues in terms of reference for EIAs.

(3) To inform EIA practitioners of the importance of the health component of EIA by:
— issuing practical guidance in handbooks, manuals, etc.
— training courses, conferences, seminars
— improving access to relevant information.

(4) To inform the general public and the media of the value of EIA (with a strengthened health component) in maintaining and protecting health through:
— popularising publications on EIA and environmental health
— documentary television programmes on health effects of development
— collaboration with non-governmental organisations in conferences, training and other activities
— education and communication of the concepts of comparative risks

In implementing this objective of increasing awareness of the value of EIA as a means of protecting environmental health, consideration must be given to the most appropriate medium of communication for each audience. The second objective is:

● *to encourage the transfer of knowledge, experience and expertise from relevant health professionals to the EIA community*

(1) Contract a major review of the applicability of principles and practice of public health, toxicology and epidemiology in EIA.

(2) Develop training programmes, simulation games and case studies to demonstrate the application of analytical capabilities of health disciplines in EIA.

(3) Support symposia, workshops and study tours for health professionals and EIA communities to exchange ideas, experience and knowledge.

(4) Identify projects as 'pilot studies' on the integration of health disciplines into EIA.

The third objective is:

● *to improve practice in EIA with respect to the assessment of health impacts*

(1) Determination of the scope of an EIA should always involve consultation with health professionals.

(2) All aspects of the project, policy or plan should be described and assessed with regard to their potential health effects, for example:

— raw materials, intermediates, products, by-products and waste products

— identification of potential accidents and routine but intermittent emissions

— handling of hazardous materials

(3) Information on the environment should include base-line health data such as:

— Health for All indicators

— lifestyle and behavioural characteristics which might affect susceptibility to health impacts

(4) To ensure that potential health impacts are identified, multi-disciplinary EIA teams must, at least, consult with health professionals. Identification of health effects may be facilitated by specific reference to health impacts in checklists and matrices, and following through cause–effect relationships to health effects. In addition to direct health effects, examples of health impacts may include:

— immigrant workers introducing new infectious diseases to an area or vice versa

— increase in population at risk in the vicinity of a hazardous installation

— reduced disease from improved waste management

(5) Where potential health impacts have been identified every effort should be made to predict the nature and extent of those effects in quantitative terms, while drawing attention to uncertainty in levels of

prediction and probability of occurrence. Approaches to prediction may include:
— use of exposure models
— use of dose–effect relationships
— use and interpretation of epidemiological and toxicological data
— comparison with similar health problems elsewhere
— drawing on expert judgement of health professionals

(6) A range of criteria are available for evaluating health effects. An EIA may refer to, among other things:
— environmental quality standards established for the protection of health, for example, air, water, food stuffs, noise
— occupational exposure limits with appropriate safety factors
— acceptable daily intakes
— early warning standards for public health

Where no limits or standards exist, the EIA team may use established risk levels, for example, negligible effect levels, virtually safe levels, etc., to interpret results for decision-makers. To assist interpretation, the EIA team should also present:

— available information on public perception of health risks
— comparative risks
— estimate of monetary cost of health effects such as costs of additional health care.

(7) If evaluation indicates that health effects are unacceptable, measures should be taken to mitigate impacts. In addition to mitigation measures traditionally considered in EIA, for example re-siting and pollution control technology, measures to mitigate health effects may include:
— improved provision of health services and infrastructure
— public health education
— improved training of health personnel
— emergency response plans
— measures to change diet and behaviour
— supplementing diet to counteract deficiencies

(8) Where health effects are key issues, appropriate health indicators should be monitored during construction, operation and, if necessary, after decommissioning. Contingency plans should describe appropriate action, if monitoring indicates that health effects have reached unacceptable levels.

7.9 THE ROLE OF AUDITING AND MONITORING IN EIA

7.9.1 Introduction

EIA is a predictive process. Through the application of a structured approach to the collection and analysis of data, its aim is to identify and assess the likely environmental consequences of pursuing a pre-determined course or set of actions, be they individual projects, development programmes or policies.

In this way, it contributes to the collection of information on EIA and EHIA, and to the formulation of environmentally sensible decisions. There is less confidence however, in the ability of EIA to predict environmental change with the required level of accuracy, and to successfully identify and implement appropriate measures to manage environmental change once it has occurred. This has resulted in increased research activity in Europe and North America into the predictive accuracy of EIA and EHIA, and the effectiveness of measures to manage environmental change. This section outlines the role of monitoring and audit and how it might be incorporated in the current assessment and management process.

7.9.2 Definitions

Various terms for monitoring and audit and related activities are in use. However, to avoid confusion there is a need to identify and define the nature of follow-up activities.

Post-project analysis has been adopted as the general term used for research and supporting activities which take place subsequent to project construction. It can be distinguished from project implementation which provides a control function rather than one of review and which operated within an immediate as opposed to long-term time horizon.

Monitoring, in general terms, is the systematic collection of data through a series of repetitive measurements. A number of different monitoring activities are relevant to EIA. They are:

- *Base-line monitoring* which refers to the measurement of environmental parameters during a representative 'pre-project' period in an attempt to determine the nature and ranges of natural variation, and where appropriate, to establish the process of change.
- *Effects or impact monitoring* involves the measurement of parameters during project construction and implementation in order to detect environmental change which may have occurred as a result of the project.
- *Compliance monitoring*, unlike the previous activities, is not directed at environmental parameters, but takes the form of periodic sampling

and/or continuous measurements of levels of waste discharge, noise or similar emission or introduction to ensure that conditions are observed and standards met.

7.9.3 *Audit* **is the term borrowed from financial accounting to infer the notion of verification of practice and certification of data. In EIA, the term refers to:**

(i)   the organisation of monitoring data to establish the record of change associated with a project; and

(ii) the comparison of actual and predicted impacts for the purposes of assessing the accuracy of predictions and the effectiveness of impact management practices and procedures.

On the basis of monitoring data, auditing examines the reasons for variance between actual effects and predictions. When considering in the context of the nature and quality of the data base and the choice and application of assessment and mitigation methods, this analysis contributes to scientific understanding of impact definition and prediction and EIA methodology. Clearly, it involves considerable interpretation and judgement and extends beyond the more narrow notion of audit as an activity implemented for the purposes of ensuring compliance with some form of predetermined objective or standard.

7.9.4 The Rationale for Audit

In the absence of follow-up, EIA is a linear process without scope for incorporating experience generated by one project into the assessment and management of another. Actual changes in ecological and social systems which occur as a result of a course of action are rarely related to the anticipated effects of that action, and so there has been a tendency to duplicate research and generate unnecessary information for each new project and each new assessment. The result has been to constrain the activity of the 'learning experience'. Equally, the notion that projects may be regarded as experiments, to determine the scientific accuracy of impact assessment and the effectiveness of management activities, has had little support or encouragement.

The development of a post-decision framework of analysis as a 'back-end' to the process of EIA is an emerging research focus. The objective is to define the mechanism whereby information derived from monitoring can feedback into the EIA process to achieve improvements in the way in which project impacts are identified and assessed, and how assessment and management as a process operates.

Feedback, therefore, builds continuity into the process, between the pre- and post-decision phases of the project cycle. It extends the conventional understanding of EIA beyond the decision point to incorporate management of project effects through monitoring and the measurement of performance, through the assessment and evaluation of predictive accuracy and management effectiveness.

The main point to be emphasised, is that monitoring and audit are critical if the concept of integrated and adaptive EIA is to be placed in an operational context, where scientific knowledge of ecological and social processes is characterised by uncertainty and where project induced change in complex natural systems is predicted, at best with difficulty. Coping with the unexpected requires a flexible approach in which monitoring, supported by audit is part of the continuing process of experimental design and adaptation.

7.9.5 Monitoring and Audit in the Context of EIA

The conventional understanding of EIA is as a process which identifies and assesses the nature of possible environmental change. A fundamental component of the process is monitoring. As described above, a range of activities can be considered which are often regarded, in practice, as separate and independent of each other. This approach places severe constraints on the continuity or monitoring and its ability to link environmental change to project activities.

Notionally, monitoring should be considered a continuous process, integrated with that of EIA. Base-line monitoring should be structured by initial screening and scoping activities and subsequently refocussed at the stage of impact analysis. It should continue through implementation and operation and by providing data about project induced changes, enable the readjustment or 'fine-tuning' of impact management measures. This in turn triggers the audit process which establishes the feedback link with impact prediction and management.

In this feedback process, it is useful to distinguish between two objectives, impact mitigation and process development. The first relates to the utility of data in correcting and adjusting mitigation measures, and project design and implementation activities. The second relates to the utility of experience generated to improve the practice of environmental assessment and management. The difference between the two is a function of the point in time at which the information generated by monitoring is relevant. Generally, monitoring for the purposes of impact mitigation and management is 'action specific' and 'user orientated', whereas process development has wider application and exists within broader time horizons. Altogether, however, follow up activities combine to refine the assessment and management of future projects as well as contribute towards advancements in the 'state of the art' of EIA.

7.9.6 Criteria for Application of Audit
The accuracy of prediction amongst other things is a function of the quality of information and capability of tools available. Accordingly, the degree of certainty with which predictions are made is likely to vary considerably between specific impacts and projects.

Various approaches to audit can be identified, emphasising different elements of the process. Some differ in their objective whilst others differ in their approach and the manner in which data is used. Few studies of the scientific/technical type have been undertaken in comparison with the number of EIAs that have been prepared since 1970. It is not possible, therefore, to make a retrospective evaluation of their contribution to the evolution of environmental planning and management. However, a small number of studies have been undertaken in which certain aspects of EIA have been examined.

7.9.7 Lessons Learned from Audit Studies
These and other audit studies have made an important contribution to knowledge about the effectiveness of EIA in achieving more environmentally acceptable decisions and development initiatives. In particular, four areas can be identified in which their contribution is especially appropriate.

7.9.8 Base-Line Studies
Historically, base-line studies have described the environmental setting of the development, emphasis being placed on the production of a comprehensive listing of environmental characteristics rather than determining the variation of these characteristics both spatially and over time. Audit studies have suggested that there is little integration between the acquisition of base-line data and the prediction of potential impacts. Clearly, as has been stated earlier both activities should positively reinforce each other. Refining the nature of the impact predictions enables the scope and nature of base-line studies to be adapted as necessary.

7.9.9 Impact Prediction
Whilst impact prediction is perhaps the fundamental element of EIA, the manner in which predictions are communicated represents a major

limitation to the utility of EIA. The following weaknesses have been recognised by a number of audit studies:

- poor identification of those variables subject to an impact
- failure to predict the magnitude of impact (e.g. size of population affected, area of land disturbed, increase in noise levels expected)
- failure to describe impacts in spatial or temporal terms
- failure to make a probabilistic estimate of the likelihood of an impact
- the significance of impact is often poorly defined
- confidence limits, based on the quality and quantity of information available are seldom stated

There is a clear need for impact predictions to be stated as *impact hypotheses*, so that they may be statistically tested on the basis of monitoring data. For this to be possible, the assumptions on which predictions are based must be clearly stated.

7.9.10 Impact Monitoring

As described earlier, the role of monitoring is perceived as the identification of variation in environmental parameters which can be attributed to the presence of development and not to some additional factor. However, audit studies conducted have shown a common failure to conduct monitoring in a manner that allows either impacts to be identified or casual relationships to be established. Clearly, monitoring programmes must reflect impact predictions made during the assessment if these are to be verified and experience generated. For example, there is little value in monitoring ambient day time noise levels in nearby residential areas if a prediction was defined in terms of night time levels at 100 m from the perimeter of the development.

7.9.11 Format of Environmental Document

To be instrumental in influencing development decisions, environmental documentation must be submitted early on in the project cycle. Paradoxically, the information on which the assessment is based, is often incomplete.

The development of a project is a dynamic exercise, whereas an environmental document, by necessity, can only describe a situation at a single point in time. Of greater value, therefore, would be a dynamic document which facilitates through, for example, a loose-leaf format, the incorporation of changes in project design and of information generated by base-line studies. In this way, the assessment process is continually updated ensuring that any impacts resulting from changes in design characteristics are identified, evaluated and if necessary managed by the implementation of mitigation measures.

7.9.12 Some General Conclusions
There are two main areas of concern for the future. First, there is a need for further research, as the observations made are based on only a few studies specifically undertaken to evaluate the utility of EIA. Many of the investigations into the environmental effects of specific projects or actions in recent years have limited value in improving EIA practice as a management tool. This is because they do not examine the role of methods and techniques, and in particular, their ability to accurately predict the nature of likely environmental impacts.

Secondly, to successfully implement audits there are two main requirements as follows:

- the environmental documentation (EIS) must include explicitly stated predictions which can be tested statistically, and
- monitoring programmes which are designed with the purpose of meeting the information needs of audit studies.

The whole EIA process must be designed with the ability to perform an audit. Auditing cannot be an exercise of 'afterthought'.

7.10 BANGLADESH CENTRE OF ADVANCED STUDIES (BCAS) ENVIRONMENT SECTOR ACTIVITIES

7.10.1 Introduction
The Bangladesh Centre of Advanced Studies (BCAS) is a private non-profit making scientific research institute with a major programme on Natural Resource Management Environment and Development. It has undertaken a number of activities and projects under this programme. Some of these are listed below.

7.10.2 Environmental Aspects of Agricultural Development
A workshop on Environmental Aspects of Agricultural Development in Bangladesh organised by BCAS in 1986, brought together scientists and experts from different fields including natural resources, agriculture, engineering, health and social sciences.

A wide variety of issues were discussed and about twenty problems and key questions were identified and presented at the final session. The Minister of Agriculture endorsed the proposals and requested detailed research plans to be submitted. The research proposals were expanded into three research projects which were undertaken by BCAS with funding from the Ford Foundation. They are:

- environmental effects of ground water utilisation;
- environmental impacts of shrimp versus paddy cultivation in the coastal districts of Bangladesh;
- agricultural developments with respect to nutrition with special emphasis on the role of women.

7.10.3 Environmental Aspects of Surface Water Systems

A similar workshop on Environmental Aspects of Surface Water Systems of Bangladesh was organised by BCAS and produced a series of research problems and research questions that needed to be answered in the area of surface water systems. The Minister of Water Resources endorsed the recommended research questions and expressed support for any projects that are prepared from the recommendations.

A team of experts subsequently developed three major research projects which were undertaken by BCAS with funding from the Ford Foundation. The projects are:

- environmental impacts of selected flood control and drainage projects;
- study of haor and pond ecosystems in Bangladesh;
- environmental effects of industrial, urban and agricultural pollution.

7.10.4 National Conservation Strategy (NCS)

A conference on conservation and development held jointly with the International Union for Conservation of Nature and Natural Resources in 1986, recommended that a National Conservation Strategy for Bangladesh be prepared.

The Government of Bangladesh subsequently constituted an 18 member inter-ministerial committee chaired by the Minister for Agriculture which recommended that a prospectus for a National Conservation Strategy be prepared. The Draft Prospectus was completed in 1987 with financial input and a consultant from ICUN. The Prospectus identified the priority areas of concern and developed a methodology for carrying out the full NCS process including all the sectors and their interactions to be done in Phase 2.

7.10.5 Social Forestry

BCAS has a major interest in the forestry sector of Bangladesh which is an important resource of the country and in danger of rapid depletion. One area of promise in maintaining and planting of more trees is in the social forestry sector which includes both government agroforestry, non-government organisations tree planting programmes and private, homestead tree planting. Accordingly BCAS organised a workshop of key people and others involved in social forestry activities in 1986. The major issues and problems were discussed and future areas of research were identified. This programme is being carried out in collaboration with the Department of Resource Management, University of California, Berkeley, CA. BCAS are currently carrying out the following research projects under this programme.

- community forestry practices and programmes in north western Bangladesh;
- women and social forestry, case studies from Bangladesh villages;
- tree tenure systems in homestead and farmland of Sreepur.
- socio-economic and cultural aspects of trees in rural households;
- forest and land/tree tenure laws in Bangladesh, myth and reality.

7.10.6 Fisheries

Fisheries, both inland and marine, is a major natural resource of Bangladesh. It provides a major source of protein for the vast majority of people and accounts for a large part of rural employment. It is an area which has been relatively ignored in terms of research, building up of a data base and resource inputs. Fisheries have been identified by BCAS as an important and neglected resource and is building up a data base on the country's fisheries resource as well as expertise in studying and managing the resource optimally.

Amongst the activities in this sector are the following:

- The Ministry of Fisheries and Livestock, Government of Bangladesh through the Directorate of Fisheries has appointed BCAS to carry out a monitoring and evaluation of the Governments New Fisheries Management Policy for inland fisheries. This programme will monitor twelve inland water fisheries in three ecological zones under different management types to determine the effects of different managements both on the fisheries resources and also on the socio-economic condition of the fishermen.

● BCAS carried out an analysis of the fisheries sector activities and prospects in three coastal districts in Bhola, Patuakhali and Barguna on behalf of DANIDA.

● BCAS carried out the study on the probable impacts on women of the First Aquaculture Project funded by the Asian Development Bank in 1986.

● BCAS is currently working with the Overseas Development Agency of the British Government, to carry out a survey of culture fisheries activities in north Bangladesh and is cooperating with the Faculty of. Aquaculture at the University of Stirling to develop aquaculture activities in Bangladesh.

● BCAS has a collaborative arrangement with the International Centre for Living and Aquatic Resource Management, Manila, Philippines to carry out studies on fisheries resource management in Bangladesh.

7.10.7 State of the Environment: A People's Report
One of the problems in devising development programmes or initiatives in developing countries, is to incorporate the people's opinions and perceptions into the planning process. This is also true in the field of environment. BCAS, along with about ten major national non-government organisations active in Bangladesh is carrying out an exercise to try to determine the people's own perception of their environment and its problems from the grass root level. The aim being to synthesise their opinions and perceptions through a scientific screening process and bring it to the attention of policy-planners, donors, educated persons and general public in the form of a report, popular articles and TV and radio programmes. BCAS is acting as the secretariat and implementing agency of the project which is coordinated by a committee.

The project is being funded by the national NGO sector itself and no external funding is being accepted.

REFERENCES AND BIBLIOGRAPHY
Environment Canada (1985) *Proceedings of a Conference on Canadian and International Experience in Audit and Evaluation of Environmental Assessment and Management*, Environment Canada, Ottawa, Canada.

Tomlinson, P. and Atkinson, S.F. (1987) Environmental audits: proposed terminology, *Environmental Monitoring and Assessment*, **8**, 189–198.

UNECE Group of Experts on EIA (1982) *Post Project Analysis in Environmental Impact Assessment*, Env/GE 1/Ro2 Note by Secretariat.

CEMP (1988) *Environmental Monitoring and Audit. Guidelines for Post-Project Analysis of Development Impacts and Assessment Methodology*, Draft report prepared for Environment Canada (unpublished).

Sadler, B. (1988) *The Role of Monitoring and Audit*, Paper presented at a Seminar entitled 'Environmental Impact Assessment', Cairo, Egypt.

Clark, B.D., Bisset, R. and Tomlinson, P. (1983) *Post Development Audits to Test the Effectiveness of Environmental Impact Prediction Methods and Techniques*, PADC Environmental Impact Assessment and Planning Unit, Department of Geography, University of Aberdeen.

Beanlands, G. and Duinker, P. (1982) *Project on the Ecological Basis for Environmental Impact Assessment in Canada: Progress Report*, Institute of Resource and Environmental Studies, Dalhousie University, Halifax, Canada.

Henderson, L.M. (1987) Difficulties in impact prediction auditing, *EIA Worldletter*, May/June, 9–12.

BP Petroleum Development UK Ltd (1986) *Guidelines for Environmental Auditing of BPPD UK Operations Facilities*, BP (NWE).

Bennett, B.L. (1986) *Value-for-Money Auditing at Ontario Hydro*, Paper presented to the Netherlands Centre for Studies.

Chapter 8

PART I: Forecasting the Vector-borne Disease Implications of Irrigation Projects: A Case Study from Africa

Techniques for predicting health impacts enable planners and designers to assess the potential health consequences of planned development at an early stage. Armed with such assessments, safeguards and mitigation measures may be incorporated into project design or operation. While there are many methods of assessing environmental impact, there remain few formal methods of assessing health impact. This paper will describe the method developed by a joint WHO/FAO/UNEP panel of experts on environmental management for vector control (PEEM). The method focuses on communicable diseases which are transmitted by the bite of insects or through water contact. It emphasises: rapidity and simplicity; intersectoral communication; and use of prior experience. It is illustrated by a case study from Africa.

8.1 INTRODUCTION

Water resource developments (WRDs) are difficult to plan and manage, requiring the coordination of a number of government sectors and the anticipation of a diverse range of problems. If health planning is to be incorporated in WRD, two basic requirements must be met. The first of these is the political will to take positive action.

The second requirement is a rapid, cheap and simple procedure for determining whether and how the first requirement (for safeguarding health) can be fulfilled. The procedure should be widely available to those without specialist knowledge of health. Community groups will need to apply it. So will specialists in other disciplines: notably the engineers, economists, politicians and agriculturalists, who are often the principal decision-makers in the water development drama.

Health issues appear to have been neglected in many development projects. Yet a recent survey of engineers ranked health fourth in priority out of 43 research needs, in connection with Third World irrigation projects (Abernathy and Pearce, 1987). A distinction was made between the need for more scientific research and the need to crystallise existing knowledge and to improve procedures for its dissemination.

In 1981 a joint WHO/FAO/UNEP panel of experts on environmental management for vector control (PEEM) was formed to address these issues. One of PEEM's initiatives is particularly relevant to the planning process: the publication of a guidelines series. Volume 1 was a Guideline for incorporating health safeguards into irrigation projects through intersectoral cooperation (Tiffen, 1989). Volume 2 was a Guideline for rapidly assessing the vector-borne disease implications of water resource development (Birley, 1989).

Donor agencies are now increasingly recognising the need to incorporate health impact appraisal into project planning. During 1990, for example, the Overseas Development Administration of the British Government funded a major new initiative at the Liverpool School of Tropical Medicine. The new programme has considerable funds to help developing countries to make appraisals of health impact and also to train nationals.

8.2 THE PROJECT CYCLE

Before designing an assessment procedure it is important to consider in more detail how and when it should be used within the project cycle.

The project cycle for a formal WRD starts with the identification of potential projects by negotiations between donor and borrower agencies. There follow a series of reports of increasing complexity and detail referred to as feasibility studies. These lead, finally, to project implementation, construction and operation, which may be accompanied by evaluation and monitoring. After many years of operation a new cycle may commence with a proposal for rehabilitation.

Environmental health impact assessment should form one component of the feasibility study. It is an activity which should not be confused with base-line surveys. In general, there will be neither the time, the staff nor the financial resources to undertake a detailed scientific survey of all the

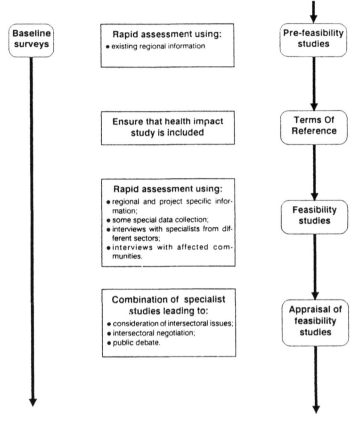

Fig. 8.1. Health impact assessment and the project cycle.

components of the natural environment and human community which interact to create a potential health hazard. Development disasters worldwide attest to the simple but sad fact that too little is done too late. Environmental health impact assessment in all but the richest countries must, then, concern itself with the art of the possible: forming rapid assessments of key issues with a minimum of reliable data supplemented by a maximum of knowledge and experience (see Fig. 8.1).

In the formal terms of the project cycle, health impact assessment starts with a rapid survey of existing regional information. Questions to ask during this pre-feasibility study include: which vector-borne diseases occur in the region; what is their relationship to water; what is the capacity of the existing health service? The results of the pre-feasibility study are used to specify terms of reference (TOR) for the more detailed feasibility study. It is vitally important that at least one among the many specialist studies which are specified by the TOR is explicitly concerned with public health.

The feasibility study will still depend on limited and inadequate data but regional information will now be supplemented by more project specific information. There should be an opportunity for site visits and interviews with specialists from different sectors and representatives from affected communities. The PEEM Guidelines suggest which questions should be asked at this stage.

Finally, each specialist group will submit a report of its assessment study to a coordinating committee. It is the task of this committee to appraise the reports and to determine where intersectoral linkages are weak and safeguards or mitigation measures are needed.

8.3 DESCRIPTION OF THE ASSESSMENT PROBLEM

Birley (1985, 1988) has described the assessment dilemma in some detail. It consists of structuring a complex problem into smaller and more manageable components and then considering the interactions between these components. The general methodology employed is referred to as environmental impact assessment or EIA. This may be defined as follows (PADC, 1983):

any activity designed to identify and predict the impact on the biogeographical environment and on people's health and well-being of legislative proposals, policies, programmes, projects and operational procedures and to interpret and communicate information about the impacts.

The problem may be structured by distinguishing three main sub-components: community vulnerability; environmental receptivity; and the vigilance of health services (these terms were borrowed from the 1966 WHO Expert Committee Report on Malaria).

8.3.1 Community Vulnerability
The human community associated with WRDs consists of a number of distinct groups such as current occupants, scheduled migrants, un-scheduled migrants, relocatees and temporary residents (see Table 8.1). The susceptibility of each human group will vary according to the degree of prior exposure to each disease and according to their general state of health and well-being.

People who are displaced from an environment with which they are familiar are subject to stresses which affect their well-being in many

Table 8.1

A simple classification of human groups associated with resettlement
projects

Scheduled migrants	Settlers selected by government
Unscheduled migrants	Self-selected settlers, squatters and encroachers
Relocatees or evacuees	Communities displaced by the project
Temporary residents	Construction workers and seasonal farm labourers

complex ways. For example, they may be short of food until the first harvest
can be gathered. Indicators of well-being can be listed (PEEM volume 2).
Contact with vectors or infected water sources will depend on the siting of
settlements, provision of domestic water supplies and sanitation and the
nature of work and social life.

Each construction worker may attract as many as 10 temporary residents
who will supply goods and services and live in unplanned settlements.

Scheduled migrants may be chosen from the young and most fertile
segment of the population and the age structure of the settlers will diverge
from the age structure of the larger community.

Example
Planners may concentrate on resettlers when a valley is flooded. However,
a fishing community may be attracted by the abundance of fish in the new
reservoir. Woodcutters may be attracted by the new communication route
allowing access to the hinterland. Such unscheduled migrants may come
from distant river basins, bringing new parasites and living in unsanitary
conditions. The planning document may ignore them completely.

8.3.2 Environmental Receptivity
WRDs can substantially alter the local environment thus introducing new
disease vectors and parasite reservoirs. The creation of large expanses of
open water promotes the growth of aquatic and terrestrial vegetation,
alters ground water levels and attracts animals or birds. The forecast must
indicate the main species of vectors, animals and plants which will be
encouraged or discouraged by the project. The effect of human activity on
the environment may be classified (PEEM volume 2). Broadly speaking, if
a vector occurs in the region and a favourable breeding site is created in the
WRD then, sooner or later, that vector will colonize the site.

An abundant vector community is only important if it has contact with
human beings. Contact between people and vectors may be increased or

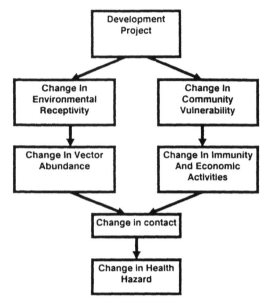

Fig. 8.2. How water development can affect human health.

decreased by changes in behaviour of either vectors or people, or by changes in vector abundance. Each species of mosquito has favourite locations and times for seeking a blood-meal. For example, human contact with malaria mosquitoes in Southeast Asia is often restricted to the forest and the night.

The project should be designed so as to minimise contact with unsafe water. Contact between people and water containing intermediate hosts may be increased or decreased by changes in human behaviour or changes in abundance of the intermediate host. In a hot climate young children will bathe in water and consequently may increase the transmission of schistosomiasis. The contact-promoting behaviour of vectors and people can be categorised (PEEM volume 2).

Example question from the Guidelines
Could a new species of vector colonise the site from elsewhere?

Figure 8.2 suggests how changes in community vulnerability interact with changes in environmental receptivity to produce a potential health hazard. There are changes to the ecological environment which affect the number of vector species and the size of vector populations; and there are

Table 8.2
Some diseases are able to spread rapidly through a community and
require only a few contacts with infected vectors or water. Other
diseases require prolonged and repeated exposure so that the parasite
burden accumulates in the person until they eventually feel unwell or
their immunity is impaired

Fast diseases	*Slow diseases*
Malaria	Schistosomiasis
Leishmaniasis	Filariasis
Dengue	DHF?
J.E.	

changes to the human population which affect susceptibility, exposure and
prevalence.

8.3.3 Vigilance of Health Services
Health departments are often the last to hear that a WRD is planned.
Consequently, there may be no health provision in resettlement villages or
construction sites. Existing facilities in a previously underpopulated district
may be stretched by the arrival of migrants without any advance planning
of drug supplies or personnel. There may be no health inspectors available
to advise on sanitation and no spray teams to protect temporary housing.
Most important of all, there may be no environmental safeguards or
mitigation measures incorporated in the project plans.

8.3.4 Bounding the Forecast In Time and Space
The extent of the potential health problem may be bounded in time and in
space. Three phases of a WRD may be identified: pre-scheme; construction; and operation. Vector-borne diseases may be categorised
according to the rate at which disease manifestations occur in the human
community who are exposed to infection. Some vector-borne pathogens,
such as malaria and arboviruses, may have a significant, immediate effect
on the health of the community. Other diseases, such as schistosomiasis
and filariasis, depend on a gradual build-up of vectors and intensities of
infection over many years (see Table 8.2). Slow diseases tend to be caused
by helminths which do not proliferate in the human host. Intermittent
chemotherapy combined with environmental management can be used to
reduce the level of worm build-up (Bradley, unpublished). Fast diseases
tend to be caused by protozoa and viruses and the pathogen proliferates

Table 8.3
The flight range of vectors (km). Migratory flights are often aided by prevailing winds and occasionally much longer flights have been recorded. Local movement is indicated as a guide to settlement siting. Where a range is indicated, the majority of vectors will only travel the shorter distance

Vector	Local movement	Migration
Simuliid blackflies	4–10	400
Anopheline mosquitoes	1·5–2·0	50
Culicine mosquitoes	0·1–8·0	50
Tsetse	2·0–4·0	10
Phlebotomine sandflies	0·05–0·5	1

within the human host. Control is aimed at reducing incidence and attacks of the disease may remain severe.

8.3.5 Map Overlays
None of the diseases occur throughout the world and regional distribution maps provide a preliminary forecast. In Southeast Asia, for example, *Schistosoma japonicum* is largely confined to the east of Wallace's line while Japanese Encephalitis is largely confined to the west. *Brugia malayi* occurs south of the Kra Isthmus while the distribution of *Wuchereria bancrofti* is patchy but more common to the east.

In the Eastern Mediterranean Region schistosomiasis is of major health importance between Egypt and Somalia, of comparatively low importance between Syria and Saudi Arabia and of extremely low significance in most other countries (Rathor, 1987).

Within each region the distribution of diseases and vectors is limited by geophysical, biological or cultural variables. For example, many parasites cannot develop in the vector at cool, high altitudes while others are associated with specific humidity zones. In semi-arid regions the distribution may be extremely focal in relation to open water. It may be possible to overlay larger scale maps with zones of known disease prevalence. The relative position of the project site with respect to the prevalence zones may then indicate whether there is a health hazard. However, vectors have considerable power of dispersal and readily colonise newly created habitats (see Table 8.3).

8.4 FORMAT OF THE SUMMARY ASSESSMENT

Figure 8.3 and the accompanying text illustrate the kind of summary assessment that the PEEM Guidelines seek to provide. The three elements of vulnerability, receptivity and vigilance are scored using simple terms such as high, low, increasing and good. The final health hazard, for each disease and each project phase, is scored by combining the three elements. The written summary justifies the scores in terms of available information and necessary assumptions. The subjective nature of the procedure is acknowledged: other readers are free to re-interpret the scores as they wish until a final consensus is obtained. The scores for each of the three elements are based on a checklist of questions which are ordered in a flowchart in PEEM 2. The aim is to encompass the many broad issues which impinge on health. For example, have the planners provided adequate health care facilities for all the community groups who will be affected by the project?

Example to accompany Fig. 8.3

Summary health forecast for a commercial wheat production scheme somewhere in Africa.

(1) **Introduction**
A certain African country decided to promote commercial wheat production using irrigated agriculture. The project area was previously uninhabited bush. There were villages nearby which had created a mosaic landscape of woodland, pasture and subsistence agriculture. The scheme would use centre pivot sprinkler irrigation with 800 m arms to grow dry season wheat. Rain-fed maize would be produced in the wet season. It would be capital intensive, using high technology, and would employ only 400 workers.

(2) **Malaria**
(2.1) *Vulnerability*
Falciparum malaria was endemic and seasonal in the region and a major cause of hospital admissions. Chloroquine drug resistance was reported. A recent survey indicated widespread ignorance of the causative factors of malaria in the general population. The workforce would largely be unemployed urban workers with their families. They would be moving from an area where mosquito control was still active and drugs were readily available. They were moving to an area without mosquito control and where drugs could be difficult to obtain.

(2.2) *Vigilance*
Medical care: The nearby health clinic followed the general trend in that country of drug shortage and broken water supply. There were excellent health services offered by a mission hospital, which included a diagnostic laboratory, some 50 km from the project site. The mission hospital had not been informed in advance of the influx of additional settlers into its catchment zone. Although stretched, it was able to respond and cope with the extra workload, including a large scale extended immunization programme. The farm was expected to provide transport for its workforce to the mission hospital.

PROJECT TITLE	Wheat farm
TYPE	Commercial irrigation
LOCATION	Somewhere in Africa
DATE OF FORECAST	June 1987
PROJECT PHASE	Early operation
COMMUNITY GROUP	Workforce

DISEASE	COMMUNITY VULNERABILITY	HEALTH SERVICE VIGILANCE	ENVIRONMENTAL RECEPTIVITY	HEALTH HAZARD
Malaria	high	good curative moderate preventative	unchanged	high
Schistosomiasis	low	curative some prevention	increasing	low but increase possible
Sleeping sickness	very low	curative only	very low	very low

Fig. 8.3. Summary of health forecast.

Prevention: Residual house spraying would not be undertaken. Workers' houses were screened. Widespread confusion among managerial staff was noted concerning the role of maize fields as mosquito breeding sites. Mosquito nets were on sale in town but were considered too expensive as they cost 6 times the daily wage of a labourer.

(2.3) *Environmental receptivity*

The local anopheline vectors were mainly associated with rainpools. There was a large *Mansonia* biting nuisance in the vicinity, associated with swamplands (dambos). The large size of irrigated fields, more than 1 km in diameter, would deter mosquito movement. The use of sprinkler irrigation on sandy soils would not promote mosquito breeding. Seasonal rainpool breeding would continue. Anopheline breeding in the lined irrigation canal was not expected due to depth and water movement.

(2.4) *Summary of malaria health hazard*

The immigrant labour force would experience an initial increase in clinical malaria prevalence. The project would not enhance the abundance of malaria vectors. The company provided access to treatment and screened accommodation for its workers.

(3) **Schistosomiasis**

(3.1) *Vulnerability*

There were foci of schistosomiasis in the area with more *S. haematobium* than *S. mansoni*. Schistosomiasis was endemic in villages near small dams and water holes which were used for recreational bathing. The vulnerability of the immigrant workforce would depend on their behaviour. For example, if only the piped water supply was used for bathing they would be reasonably safe.

(3.2) *Viligance*

Medical care: Patients presenting with clinical symptoms at the mission hospital could expect efficient treatment.

8.5 CONCLUSION

Water resource developments change the environment and the human community, sometimes promoting contact with vectors and infection with pathogens. Our understanding of the dynamic processes involved is limited. We are uncertain about the relationships between ecological and social variables. Despite our uncertain knowledge, we need rapid assessments of health impact so that safeguards and mitigation measures can be incorporated in design and operation. A new assessment method has been described which uses an ordered series of questions to establish a

---

Prevention: A piped water supply was provided to the worker's settlement, so that contact with hazardous water could be avoided.

(3.3) *Environmental receptivity*
The irrigation scheme created additional standing water in the form of a large reservoir which was some kilometres from the site. The workforce need have no contact with the reservoir. A lined canal carried the irrigation water to the site. The steep sides of the canal would prevent water contact. However, additional barriers would be required at weirs. The settlement was a few hundred metres from a river where recreational bathing could be expected to provide a focus of schistosomiasis.

(3.4) *Summary of schistosomiasis hazard*
The increased density of human settlement near potential schistosomiasis foci could lead to a gradual increase of prevalance. Effective curative measures were available.

(4) **Trypanosomiasis**

(4.1) *Vulnerability*
Sporadic cases of human sleeping sickness had been reported in the past from local villages, but these had disappeared together with the wild game. Sleeping sickness would continue to be a hazard to hunters and gatherers who had close contact with host reservoirs. The workforce had no need to enter hazardous areas.

(4.2) *Vigilance*
The staff at the mission hospital had treated sporadic cases of sleeping sickness in the past and would continue to do so.

(4.3) *Environmental receptivity*
The scheme was surrounded by miombo woodland and close to an old tsetse barrier on the road to a game reserve. The tsetse barrier had recently been removed. The tsetse population has declined because wild game had been depleted by commercial poaching operations. The poaching was a response to the shortage of meat in urban markets. In future, the miombo mosaic would probably be used more extensively for cattle grazing and the tsetse fly would return. In the absence of a reservoir, host sleeping sickness would not present a problem.

(4.4) *Summary of sleeping sickness hazard*
Although the project was surrounded by miombo woodland and near to a game park, there were no tsetse flies and the game reservoir was much reduced. Therefore, the risk of sleeping sickness was very low.

basis from which impacts may be inferred. Future developments of the method will benefit from new computer technologies such as expert systems, hypertext and geographical information (Birley, 1985, 1988). Environmental health risk assessment is still more of an art than a science and a tool for decision-makers, not scientists. Its objectives are modest: to improve perception of a complex problem where decisions have profound repercussions on both the environment and the community. Any method which is simple to use in practice, improves perception and reduces unstructured guesswork should be given serious consideration. Finally, health risk assessments can only be of value if there is a serious commitment by both donor and recipient to safeguard health.

ACKNOWLEDGEMENTS

Thanks are due to the joint FAO/UNEP/WHO Panel of Experts on Environmental Management for Vector Control (PEEM) for their continued encouragement and support. The opinions expressed are the author's responsibility alone.

REFERENCES

Abernathy, C. L. and Pearce, G. R. (1987) Research needs in third world irrigation, *Proceedings of a Colloquium at Hydraulics Research Wallingford*, Hydraulics Research Ltd., Wallingford, UK.

Birley, M. H. (1985) Forecasting the vector-borne disease implications of water development. *Parasitology Today*, 1, 34–6.

Birley, M. H. (1988) Forecasting potential vector-borne disease problems on irrigation schemes. In: (M. W. Service ed) *Demographic Changes and Vector-Borne Diseases*. CRC Press, Boca Raton, FL.

Birley, M. H. (1989) Guidelines for forecasting the vector-borne disease implications of water resources development, *PEEM Guidelines Series* volume 2, WHO/VBC/89.6.

PADC (ed) Environmental Impact Assessment and Planning Unit (1983) *Environmental Impact Assessment*, NATO ASI Series D: *Behavioural and Social Sciences*, 14, Nijhoff, Boston, 439 pp.

Rathor, H. (1987) Predominant agricultural practices and their bearing on vector-borne disease transmission in the WHO Eastern Mediterranean region. In: *Effects of Agricultural Development on Vector-Borne Diseases*, Edited version of working papers presented at PEEM AGM. FAO, Rome, reference AGL/MISC/12/87.

Tiffen, M. (1989) Guidelines for the incorporation of health safeguards into irrigation projects through intersectoral cooperation. *PEEM Guidelines Series*, volume 1, WHO/VBC/89.5.WHO; (1966) *Expert Committee on Malaria*, 12th report, Technical report series 324.

PART II: Saguling Dam/Reservoir (Indonesia) and Nam Pong Project (Thailand)

8.7 INTRODUCTION

Environmental impact assessment (EIA) in relation to water-resource developments, involves the prediction of environmental (including social and economic) changes which are likely to affect the water resource and the surrounding environment.

The prediction of such consequences involves consideration of both physio-chemical and biological parameters. Water-resource developments have uncertain effects which may be beneficial, neutral or harmful, in both the short- and long-term.

This case study considers two EIA studies for water resource projects in SE Asia to observe how they were implemented. The brief review provides some general comments regarding the utility of EIAs and future research and 'practice' needs for EIA.

8.8 SAGULING DAM EIA

This EIA study was undertaken in Java, Indonesia, and analyses the potential environmental and social impacts of a dam and reservoir. The EIA was commissioned by the State Electricity Company and implemented by the Institute of Ecology, Padjadjaran University (1979). The dam would inundate a part of the densely populated fertile plateau of Cililin, west of the city of Bun dung. Thus, the need to analyse environmental and social

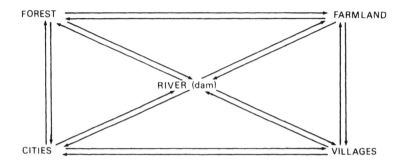

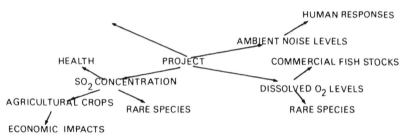

Fig. 8.4. Project and environment: holistic ecosystem perspective.

impacts was important. In addition to an account of expected impacts, the Institute of Ecology was asked to suggest mitigation measures to ameliorate adverse impacts and action to enhance positive, beneficial impacts.

The usual orientation in EIA is to take a proposed project as a starting point for analysis and to look outward from this base to ascertain how the project will affect various environmental features. This process can be shown diagramatically (see Fig. 8.4).

However, this is not a complete picture of the 'real-life' situation. In the Saguling dam case, it is true that the dam will affect a variety of environmental features in the manner shown above. Conversely, certain of the components will affect the dam/reservoir. For example, a continuation of the existing trend of reduction in forest cover might lead to increased run-off to the river and consequently the reservoir level might alter. In addition, this run-off might transport high sediment loads into the reservoir causing impacts on both dam and reservoir. Thus, the Saguling dam EIA takes a holistic ecosystem approach because it explicitly considers the dynamic multi-directional links between a project and its surroundings.

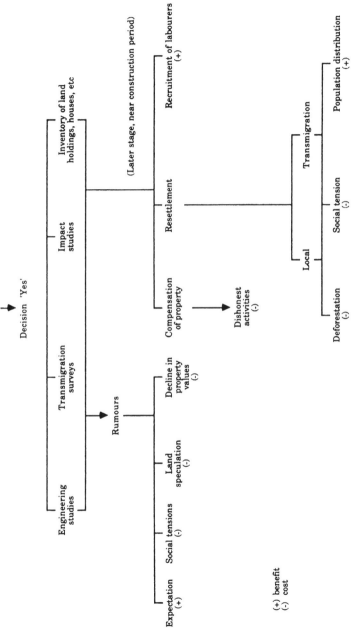

Fig. 8.5. (a) Potential environmental impact during the pre-construction period.

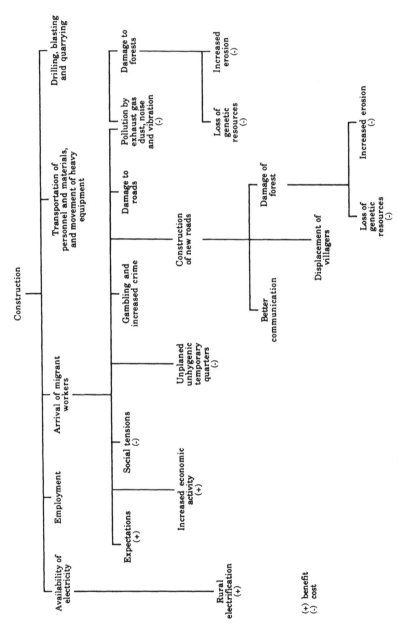

Fig. 8.5. (b) Potential impact of the construction period.

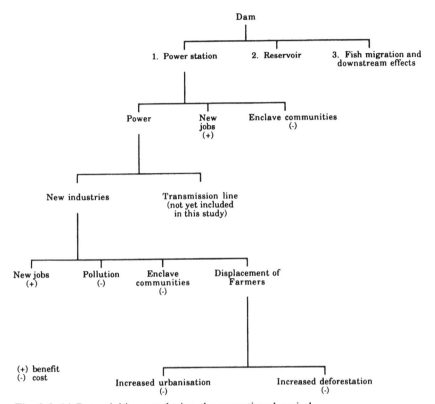

Fig. 8.5. (c) Potential impact during the operational period.

A number of EIA methods were considered before selecting a 'flow diagram' method which is a variation on a network. A number of these diagrams were constructed for various aspects of the study. Initially, three basic flow diagrams were constructed for the pre-construction, construction and operational periods (Figs. 8.5(a)–(c)).

The flow diagram for operational impacts was extended by preparing two comprehensive additional diagrams for the impacts of the reservoir in the vicinity of the dam and the downstream impacts. The EIA contains a description and account of the impacts identified in the flow diagrams. These flow diagrams not only display the impacts of the dam/reservoir on the environment, but also the effect of existing environmental trends on the dam/reservoir (Fig. 8.6). For example, population growth was thought to be a causal factor which would result in increased urbanisation, waste disposal and deforestation. The combined effects of these trends would be

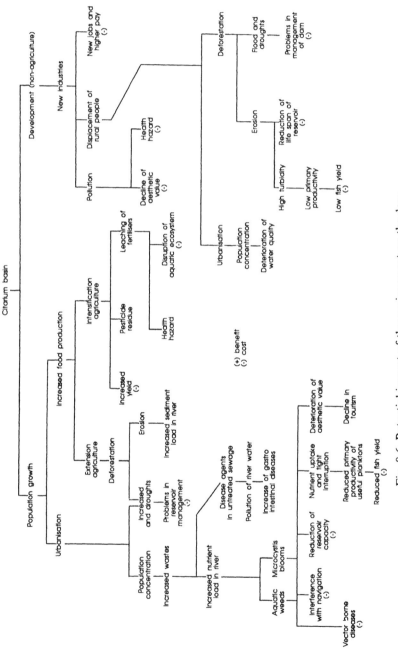

Fig. 8.6. Potential impact of the environment on the dam.

to reduce the life of the Saguling dam/reservoir because of factors such as increased sedimentation in the reservoir (Fig. 8.6).

The main conclusion reached, as a result of the construction of the flow diagrams and assessment of specific impacts, indicates that the ecosystem of the watershed was deteriorating through erosion resulting in loss of nutrients. This situation was thought likely to continue whether or not the dam was built. However, the situation was considered to be reversible and that the dam could accelerate the process. Also this process itself, could affect the functioning of the dam/reservoir.

This understanding of the relationship between the project and its environment provides the background and reason for many of the mitigation measures proposed for the project. These are aimed at not only alleviating the direct effects of the dam on local people and environmental features, but also at improving the conditions of the ecosystem and ensuring that the project fulfills its objectives. This shows the utility of EIA is not limited to achieving development at least environmental cost, although this is an important objective. EIA can also improve existing environmental conditions and aid project operation through project management. The ability of the Saguling dam EIA to identify the problems and suggest management strategies is, in considerable part, due to the 'holistic' ecosystem approach combined with the use of the flow diagrams to identify the chains of linked impacts.

8.9 THE NAM PONG DAM/RESERVOIR PROJECT

The Nam Pong project in north-east Thailand is a multi-purpose water resource development designed for irrigation (53 000 ha), power generation, reservoir fisheries, flood control and recreation. Information on some of the impacts of the dam/reservoir became available in the operational phase of the project. It was found that several adverse effects were occurring which might, in the future, affect the resources of the Nam Pong basin and the 'quality of life' of the local people. Accordingly the Mekong Secretariat of the UN Economic and Social Commission for Asia and the Pacific launched the Nam Pong Environmental Management Research Project which, implemented in three phases, aimed to provide a comprehensive assessment of the actual effects of the dam/reservoir and of likely future impacts, resulting from possible management strategies. The three phases were:

(i)   an analysis of the actual impacts (mostly socio-economic) of the project;

(ii) a base-line study of the entire Nam Pong basin covering thirteen different topics such as hydrology, limnology, fisheries, water quality and socio-economic characteristics;

(iii) integration of information from (i) and (ii) and development of a computerised simulation model to aid decision-makers manage the project in the future.

The Nam Pong model was built over a seven month period, beginning with a four week workshop to formulate the basic concepts of the model and to train Thai scientists who would continue to refine the model after the workshop. In the workshop the first task was to define the boundaries of the study. First, management actions which might affect the project in the future were defined along with indicators to show the consequences of various management actions. Values for the indicators would form the output of the model. Secondly, the spatial and temporal dimensions of the system(s) to be modelled were defined. Six functionally distinct areas were identified:

(i)   a watershed including forests and upland agriculture;

(ii) a reservoir including the drawdown zone and fisheries;

(iii) a resettlement area;

(iv) an urban area around Khon Kaen (the only large town);

(v)   further potentially irrigable areas.

The model was designed to simulate one year at a time, up to 20 years.

Within the 'parent' model a number of sub-system models were constructed. To ensure that the 'output' of each sub-system model was in a form that could be used by other sub-systems and the 'parent' model, those involved with specific sub-system models were instructed as to the type of information they had to produce as inputs to the other models. For example, the socio-economic sub-model required data on water demand for irrigation (from the land-use sub-model) to enable the calculation of crop production to be carried out.

The specific objectives of the socio-economic sub-model were:

(i)   provide indicators of the social and economic welfare of the people of the Nam Pong basin;

(ii) provide demographic and economic data needed by other sub-models;

(iii) simulate the dynamic and interdependent relationships between economic and demographic variables and the physical and biological environment within the basin (as described in the water, fisheries and land use models);

(iv) predict the level and distribution of the population and net income per capita, during the next 20 years.

Using data and certain assumptions on existing characteristics and trends, the behaviour of the two main indicators of the socio-economic sub-model, population and income were calculated for a 20 year period into the future. This was called the basic scenario. Also, since the Thai government has the objective of reducing the birthrate to 2% per year, an alternative scenario, showing the effects of this additional factor on per capita incomes, was developed.

The scenario developed from a 2% birthrate resulted in higher per capita incomes.

The model(s) can be used to show the effects of varying alternative management actions and can be used to show, quickly, the implications of changes in assumptions or the acquisition of new information on environmental and socio-economic trends. A sensitivity analysis, which would be laborious if implemented manually, can be easily undertaken by a computer. This can be achieved by incorporating in the model various differing assumptions about interactions in the model. For example, a model using average river flows will not be able to account for the effects of a flow which only occurs once in a year. Models can deal with this degree of variety if frequency distributions for all input variables were to be incorporated explicitly. With a computer-based model calculation of indicator values for various levels of input variables is fairly easy. For example, it might be that the effect of a dam on a deer population might only cause a small decrease in numbers for all values of the variables in the model. In that case no further work would need to be done. Similarly, if the impact were large for all values, in terms of population reduction, then it would be necessary to consider this an important impact requiring mitigation.

8.10 ADVANTAGES/DISADVANTAGES OF SIMULATION MODELLING

Simulation models are useful management tools to be used after a project becomes operational. Their utility does not cease with the assessment phase. The usefulness of many other EIA methods declines once a decision has been made. Should monitoring indicate detrimental impacts occurring from an operational project, then the implications of various strategies to prevent or ameliorate the impacts can be easily seen from the model. Moreover, such occurences will indicate a weakness in the model and aid its improvement and the ability of the model to predict more accurately in the future.

This method shows a tendency to concentrate on the insights into environmental problems provided by workshop discussion. In addition, it can encourage cooperation between EIA personnel and decision-makers and be sparing in its use of base-line environmental data. While these aspects are important and valuable, EIA is about predictions, and judgement on this matter needs to await information on the performance of operational models. It is hoped that such information will be forthcoming. The predictions made for alternative management strategies for Nam Pong would be a useful starting-point for an assessment of the performance of simulation models.

The application of this method is biased towards the management of resources, for example forests or particular economically important species such as salmon. Large-scale water projects have also been assessed, but few, if any, applications of assessing the environmental impacts of developments such as oil refineries, power stations and pulp mills exist. Until such uses are reported and its utility in these contexts assessed, no judgement on the wider applicability of simulation models as an EIA method can be made.

8.11 SOME CONCLUSIONS

Nearly all EIAs have been implemented for projects. However, important environmental effects arise from socio-economic changes, such as land reform, which may be directed by national or local governments. In addition, social changes can be non-directed and these also can have important environmental repercussions. The environmental effects of these social changes can be more profound and damaging than all the impacts from major projects. EIA has yet to find a role in this context. EIA can also be used at the plan, programme and policy level, but there is little experience of implementing EIA for such actions and virtually none in industrialising countries, which face many problems such as:

- lack of base-line data;
- lack of trained manpower;
- lack of finance;
- weak institutional structures.

Despite these problems, there is every reason to believe that EIAs will increase as international agencies encourage their preparation. Such agencies also have a duty and responsibility to monitor and assess the performance of EIAs which they commission or encourage. Only in this way, can the value of current practice be determined and improvements, if needed, be identified and put into effect to ensure that economic development occurs with the minimum of environmental disruption.

REFERENCES AND BIBLIOGRAPHY

Institute of Ecology, Padjadjaran University (1979) *Environmental Impact Analysis of the Saguling Dam*, Volume 11A Main Report, Institute of Ecology, Padjadjaran University, Bandung, Indonesia.

Mekong Secretariat (1981) *Nam Pong Environmental Management Research Project, Phase III*, Mekong Secretariat, UN Economic and Social Commission for Asia and the Pacific, Bangkok, Thailand.

PART III: Health Aspects in EIA of Two Industrialised Areas in Poland

8.12 INTRODUCTION

The long-range objective of this project was to assist the Polish Government in incorporating environmental protection consideration into planning and decision-making processes. The immediate objectives were:

- assessment of environmental impact by increasing the level of knowledge of the effects of pollution including health impact of development in the pilot zones as defined in the work plan;
- definition of preventive and corrective measures to eliminate or minimise the environmental impacts;
- development of an EIA system as a tool in the decision-making process.

The Institute for Environmental Development, Katowice, commonly referred to as the Environmental Pollution Abatement Centre was responsible for the research.

8.13 PRESENTATION OF TWO PILOT AREAS

To achieve the project objectives, two pilot areas were selected both with common characteristic long-term plans for rapid economic development backgrounds and levels of urban and industrial development (Fig. 8.7).

Fig. 8.7. Location of two pilot areas in Poland.

8.13.1 Pilot Area 1: South-East Edge of Upper Silesia Agglomeration (SEUSA)

The first pilot area is part of one of the oldest industrial regions in Poland where economic development is based on zinc–lead ore mining, hard coal mining and extraction of construction materials. On the base of resources, non-ferrous metallurgical industry, electrical power production and chemical industries have been developed. Part of the area is occupied by agriculture. Future development of this area is connected with hydraulic development of the Vistula-River Basin and therefore additional industrial development can be expected in the vicinity of the river. Further development of the mining, electrical power and hard coal industry is expected.

The following problems of this area were considered:

(i)   unfavourable river flow characteristics;
(ii) the effect of construction dams and reservoirs in the Vistula River to increase navigation;
(iii) increased municipal and industrial water consumption;
(iv) the effects of increased coal combustion.

8.13.2 Pilot Area 2: Legnica-Glogow Copper Mining Region (LGCMR)

The second pilot area is relatively new as an industrial region. In the 1950s after discovering large copper deposits the development of the area started and then through the comprehensive regional plan in 1960s the full development was based on industrial production, agriculture and extensive social infrastructure. After realisation of this twenty years' plan the developed region with extensive new industrial and urban structure has been created.

Future further development involving the intensification of copper ore mining and copper smelting required the detailed analysis of possibilities of additional economic growth. This development will cause the following problems:

(i)   use of area for the storage of wastes from the metal recovery and enrichment process;
(ii) increased salinity of mining wastes;
(iii) increased emission of heavy metals with adverse impacts on human health, air, water, land, agriculture.

In both areas, human health was the common problem which arose as a consequence of further development.

8.14 APPROACH TO ENVIRONMENTAL HEALTH IMPACT ASSESSMENT

EIA is concerned with disturbances caused by human activities in the environment. The problem is complicated because people cannot stop their damaging activities, and can only seek to minimise adverse effects. It is important to know what is the worst, the most hazardous, the weakest place and weakest receptor. The latter is the human being. In all environmental considerations the human being should be the central object, since sooner or later, all negative impacts focus on the human being (Fig. 8.8). Impacts should be classified as adverse only in such cases when they lead in consequence to a human and have a negative impact on mankind.

Health is the most important element of the human structure as far as it is considered as a receptor. Other aspects are social, aesthetic, cultural and economic characteristics. In both pilot areas the environmental impact analysis has included negative effects on human health as one of the main conflicts. The general approach to health impact assessment consisted of

a AIR
b WATER
c SOLID WASTES (DUMPING)
d AREA OCCUPATION
e NOISE
f VEGETABLES

g UNDERGROUND WATER
h POTABLE WATER OR LAKE
i AGRICULTURE
j SOIL (EROSION)

k HEALTH
l RECREATION
m FAUNA
n FORESTS

o ECOSYSTEMS
p AESTHETICS
q CHANGE OF AREA
   UTILISATION

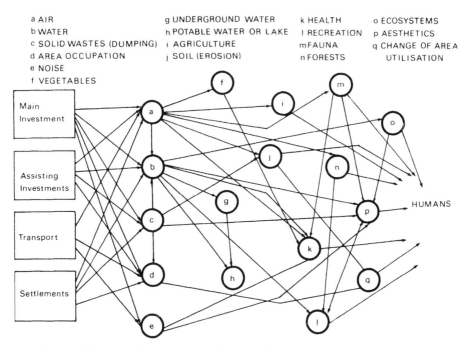

Fig. 8.8. Diagram of impacts on environmental system.

analysing these conflicts, which preceded health effects as shown in Fig. 8.8.

In the first pilot area (SEUSA), the main elements leading to negative health consequences was atmospheric air (Fig. 8.9) and food (vegetables grown in this area). As far as air pollution is concerned, dust containing heavy metals and hydrocarbons have been selected for analysis. The actual concentration of these parameters was determined as well as by calculations based on mathematical dispersion models. The areas of the highest dust concentration level were potentially the most hazardous from the food production level.

Evaluation of the harmful exposure of the SEUSA population to lead and cadmium was carried out using as reference certain of the vegetables being grown in this area. Data indicated that eating only three vegetables such as parsley, carrots and beets grown in the pilot area causes that body to have cadmium and lead quantities equivalent to 20–30% of the provisionally tolerated weekly intake PTWI of these metals. PTWI defines the quantity of a substance which can be fed to the human body over a span

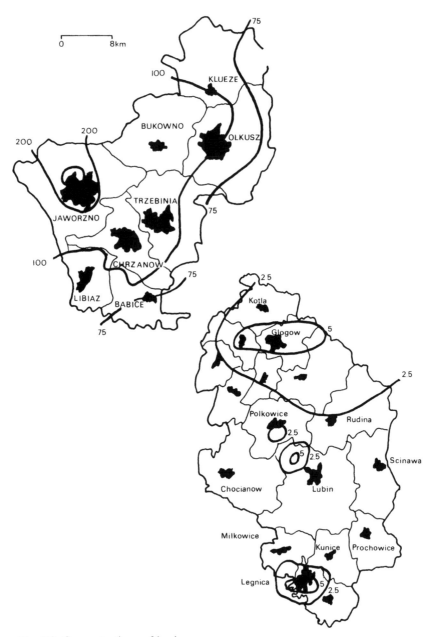

Fig. 8.9. Concentrations of lead.

of one week without detriment to health. PTWI defined by WHO/FAO for a male adult amounts to: lead 3 mg, cadmium 0.4–0.5 mg. Compared with the relatively unpolluted reference areas the degree of the pilot area's population exposure to lead is 5–8 times higher and to cadmium, 2–7 times higher.

In the second pilot area, studies on harmful health effects of human activities were carried out for every two distinct areas. First, zones surrounding two smelters 'Glogow' and 'Legnica' being under direct influence and, second the remaining area. In the zones surrounding the copper smelters the following environmental components are badly affected: water, soil and vegetation, posing their direct environmental threat to man. In order to evaluate environmental hazard and changes in human health again the most sensitive group of population was selected, i.e. children living close to the smelters and being directly affected by their impact. It was found that people living close to the smelters more often suffer from troubles in breathing, blood and ingestion systems than people living outside these zones. Also among the former, cohort deviations of biochemical and toxicological factors are observed. Medical examination of children living within the zones surrounding the smelters found more health troubles, stomachaches, vomiting tendency, bigger tonsils, and more frequent communicable diseases which spread out through droplet injection, e.g. measles, mumps, smallpox, than among children living outside the zones.

A higher level of lead concentration was found in children's blood of the former cohort than in relation to the latter one. However, in comparison to past studies 1975–1978 a lower general level of lead intoxication was observed as expressed both by lower percentage of cases with higher lead concentration and by the fact that in the examined children cohort, lead concentration level in blood determined was below the approved critical level, i.e. 4 μg/100 ml. The effects of haematological determinations in red blood cells and white corpuscles did not show occurrence of any kind of diseases in a single case and all haematological factors, in mean values, were within the acceptable standards.

The studies on biochemical factors showed higher levels of methaemoglobine, heptoglobien and ceruloplasmine and lower activity of enzymes, acid phosphatase and dehydratase delta aminolevulinic acid in children living within the direct vicinity of smelters and being directly affected in comparison to the control cohort outside these zones.

Toxicological studies show that lead and cadium concentrations in children's urine indicate some slight danger due to exposure to these metals mainly around 'Glogow' smelter. On the other hand, zinc and copper concentrations in urine do not suggest any threat caused by excessive absorption of those metals.

To conclude, the results of the study indicate that in comparison to past studies in 1975–8, some differences related to biochemical and toxicological factors are evident. Higher frequency of health troubles and diseases occurred as compared to the present health condition of children living outside the zones directly impacted by copper plants. Generally, the health situation of both adults and children living within those zones is worse than of corresponding groups living in control places outside.

Concerning the effects of toxic metals, the main agent is lead (Fig. 8.9). It is reflected in higher lead levels in biological material sampled, in concentration changes and in some biochemical factors. Also the gaseous air pollutants affect human health and make people suffer from respiratory tract diseases. Relatively low concentration levels of selected metals found in biological material and concurrently small changes of biological parameters suggest that long-term exposure may stimulate some toxicological problems particularly as far as young people are concerned. Environmental exposure of people to low concentrations of metals, presently not showing any noticeable symptoms, will lead likely to metabolic changes.

The pragmatic approach to human health exposure assessment, which could also be applied for prognosis, is based on the quantitative estimation of the amount of heavy metals getting into the body by two routes: respiratory and alimentary tracts. Such quantitative assessment was carried out for the LGCMR area. The exposure assessment by the respiratory route regarded the smallest fraction of dust (less than 20 μm) and 24 h ventilation lung capacity. The analysis gave the results for three locations in the pilot area. They were from 0.161 mg Pb/week. The exposure assessment for drinking water for 21 days gave, for the same locations, the amount of 0.532–0.420 mg Pb/week. The exposure assessment in reference to food related to some of the vegetables most commonly eaten. The weekly dose including potatoes, parsley, carrots, beets and celery amounts to 3.399 mg Pb/week, which exceeds the permissible dose of 3 mg Pb/week. The relation between the above numbers for the three locations for air, water and vegetables are 1:3:6. It means that the relation between the importance of these components as sources of exposure is also 1:3:6.

Studies and investigations concerning determination of the present state of human health exposure as well as the recognition of the future hazard connected with alternative of further development of the pilot areas gave some basis for formulating conclusions.

In the first pilot area, the forecast degree of pollution, particularly air pollution, indicates a growing threat to the health of the population. The greatest hazard occurs in the vicinity of refineries, power plants, mines and plants processing non-ferrous metal ores. A very different source of hazard to man in the SEUSA will be exposure through the alimentary tract caused by consumption of agricultural products from the affected area. Some heavy metals and carcinogenic hydrocarbons can accumulate in plants. In order to reduce the exposure through the alimentary tract it is necessary to cultivate plants with lesser ability to accumulate pollution.

In the second pilot area, the health situation of the inhabitants and exposure to environmental toxic factors is to be assessed against the background of pollution levels. It is forecast, that threats could be expected in the future only within the zones surrounding copper smelters, by some air pollutants such as SO_2, NO_x, CS_2, H_2SO_4 as well as dust, the latter because of its high content of toxic metals, mainly lead and copper. In the long run, permanent and complex impact of pollutants causing their accumulation in the environment can be viewed as a danger to the environment, and to the health of the pilot zone inhabitants. Based on forecasts of copper industry developments in that area, it is expected that by 1990 emission of pollutants will significantly increase. Thus, it seems likely that the hygenic standards will be exceeded more frequently, and on a growing scale causing deterioration of the health of the people in the region. Consequently, prophylactic examinations should include impact of gaseous and dust containing metals, air pollutants on the environment and people.

8.15 FINAL REMARKS

In addition to the initial objectives of the project, some other results have been achieved. First, the EIA approach has been introduced by the Institute of Environmental Development and the new Department of System Sciences has been established with responsibility for developing new EIA techniques and application of EIA in all cases involving alternative location strategy, production technology and pollution control. It has been found that various disciplines are necessary for a full analysis of the EIA approach, such as in the fields of philosophy, psychology, social sciences, etc. The nature of EIA requires the application of 'soft' sciences — fuzzy sets theory, multi-criteria decision-making models.

Certain activities have already taken place in preparation for the introduction of EIA procedure into the planning and decision-making process in Poland. EIA education has started and the first national EIA course has been prepared. It is foreseen that EIA will become an obligatory requirement.

REFERENCE

Environmental Impact Assessment System (1983) Final Technical Report of the Project, POL/RCE 003, Warszawa.

PART IV: Industrial Development: Iron Smelting Plants, Brazil

This case study relates to an EIS for two iron smelting plants that will produce pig-iron based on charcoal, in the southern part of Maranhao State, in Brazil. The study was based on the Federal Environmental Agency SEMA guidelines for EIA, in which the matrix method was used to identify and assess environmental impacts, by analysing social, ecological, technical and economic information related to the projects. Some comments are made about the constraints and the practical solutions for developing an EIS in a developing country.

8.16 INTRODUCTION

The EIA for the iron industrial plants was developed in line with legal requirements of the National Environment Council. Both plants are located at the southern part of Maranhao State in the municipal district of Acailandia. The site is located between the large ore deposit of iron and manganese, Serra dos Carajas, and the capital of Maranhao, Sao Luiz. The two plants are close to each other and form part of the same environmental system.

The study was developed in two basic steps, involving first, the analysis of technical information regarding the location of activities, lay-out of plants, industrial processes and production phases, including the social and economic factors involved. The ecological and socio-economic aspects of the surrounding area were also studied. The second step involved the

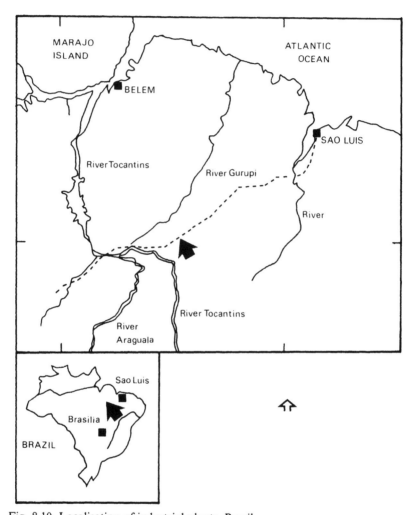

Fig. 8.10. Localisation of industrial plants, Brazil.

identification and assessment of negative (−) and positive (+) environmental impacts due to project location and operational activities, in which the matrix was used as the basic method to identify the environmental factors affected.

The systematic approach finally permitted the formulation of specific conclusions, as well as general remarks regarding the environmental programmes for both plants. These programmes related to the control and monitoring of adverse impacts, and development of beneficial impacts.

The constraints in developing an EIS in a developing country, are mainly due to the lack of information and human resources in the field of EIA.

8.17 BACKGROUND INFORMATION ON THE PROJECT

The industrial plants are part of the Great Carajas Program, PGC, coordinated by the Planning Secretariat of the Federal Government, SEPLAN. Both plants will have financial support from SEPLAN and the Superintendent for Development of the North eastern Region, SUDENE.

8.17.1 Location of the Plants
The plants will be strategically located close to the 900 km Carajas railway which connects the mining site of Carajas, Serra dos Carajas, with the port of Ponta de Madeira, near the city of Sao Luiz, capital of Maranhao State. The site is part of the district of Pequia village in the Municipality of Acailandia. (Fig. 8.10).

8.17.2 Objective of the Plants
- In general terms, both plants were designed in modules to produce pig-iron for steel plants and foundries; including the production of nodular and granular iron.
- Production is estimated at 50 000 tons per year. Future plans are to expand production up to 100 000 tons per year.
- The plants will employ approximately 200 men each, mainly from Pequia village. Workers will have special training for operation activities.
- Residential villages will be constructed for employees and dependents. These villages will have all basic infrastructures such as water and energy supply, basic sanitation, recreational areas, education, transportation and medical assistance.
- A sustainable plan will be developed for charcoal exploitation. The plan includes afforestation of adjacent areas with fast growing native species, well adapted to ecological conditions. A landscape plan will be developed for both plants.
- The project includes the preservation of approximately 360 000 m^2 of green areas. This includes the assai palm trees that occur in the Piquia river. These palm trees are important for water purification.
- The plants will have a special system of filters to control gas emissions. Industrial wastes will be managed and commercialised.

- One of the major objectives is to export pig-iron to international markets, although a great part of the production will be consumed by national steel plants.

8.17.3 Justification for Developing the Plants
- proximity (390 km) to the large deposit of ore, iron and manganese, of Serra dos Carajas;
- proximity of charcoal production area;
- availability of electric energy from Tucurui HPS;
- availability of transport system (Carajas railway, BR-222 and BR-010 motorways);
- availability of work-force at low cost;
- adequate site for installation and operation of both plants, mainly because of topographical features and environmental conditions.

8.17.4 Implantation and Operation Activities
Implantation of plants
Both plants will have almost the same type of activities for implantation: clearance of vegetation; land movements; ground level/filling construction of infrastructures (office, laboratory, toilets, charcoal and ore silos, high-oven, residential village, water supply and sewer systems, and recreational areas), and improvement of existing access roads.

8.17.5 Operation of Plants
The main operation process of each plant for pig-iron production can be divided into seven phases as follows:

- reception of raw-materials and laboratory analysis of iron ore and charcoal;
- screening of raw-materials;
- storage of raw-materials (to guarantee approximately 30 days non-stop production)
- operation of high-oven (HO): the reduction of iron ore into pig-iron is realised in the HO; temperatures are around 700°C to 1000°C. Outflow of gas is estimated in 14 000 N m^3/h. The capacity of HO is approximately 170 tons per day;
- pig-iron ingot process;
- stock of pig-iron for export;
- commercialising product.

8.17.6 Sub-Products (wastes) from the Industrial Process

- gas emisssion from high-oven (CO_2, CO, H_2 and N_2 = 14 500 N m^3/h;
- solid waste (slag) = 22 tons/day;
- fine iron-ore = 35 tons/day;
- fine charcoal = 39 m^3/day;
- cooling water = 150 m^3/day;
- noise/vibration (turbo blowers) = 90 dB (A).

8.17.7 Gas Emission Control System

The control of gases will be maintained by special equipment and by monitoring schemes. The basic equipment are primary and secondary dust collectors, gas washers, venturi washers, gas tubing security (explosion) valves. Concentration of gases after depuration will be approximately 9 mg/N m^3.

8.17.8 Noise Control

Noise and vibration from the machine will be minimised by acoustic seals. The external surrounding area of the plants will be protected against noise by tree belts.

8.17.9 Solid Wastes

All solid wastes (including fine ore and charcoal) will be sold or used as earthworks.

8.18 ENVIRONMENTAL DESCRIPTION OF THE AREA

The description of the environment (natural, social and economic) system of the (local) area of influence of the industrial plants, including general aspects of the region are given below:

8.18.1 Natural System

Geology and geomorphology

The region is classified as Itapecuru Formation (cretacean). This formation is composed basically of red sandstone, fine kaolinite and red claystone. Some spots of basal conglomerate with modified basal pebble occur very near to the project site.

In geomorphological terms, the region is predominantly formed by plains and large plateaus resulting from the South American surface.

Table 8.4
Climate

Classification of climate	Tropical (hot/humid)
Predominant wind direction	East/northeast
Wind intensity	Weak to moderate
Average annual precipitation	1·350 mm
Rainy season	December to May
Dry season	June to November
Average annual temperature	25·4 °C
Average annual evaporation	1·441 mm

Tertiary covers with erosive borders occur in the project site. The topography is slightly undulating.

Soils
The area is composed of two basic types of soils: (i) red–yellow distrofic latossoil and (ii) yellow distrofic latossoil with good drainage and permeability (PCG, 1981).

Climate
See Table 8.4.

Vegetation (IBGE, 1984)
The three types of vegetation identified are terrestrial, transition and aquatic. In the first type there is a predominance of grassland (with cattle farming activities) mixed with some spots of secondary vegetation. This vegetation is composed basically by arbustive and sub-arbustive species.

The transition and aquatic species are represented by a great collection of palm trees. The Piquia river is covered by assai palm trees, with some areas covered with aquatic (swamp) vegetation, giving a greenish appearance to the river. The river is going through an advanced eutrophication process.

Fauna
The endemic fauna is related to the surrounding vegetation and local climate. Local information indicates that the degradation of the primary forest by human activities affects the representative fauna. The group affected was mainly felines, primates and wild pig. Fauna that has adapted to new anthropological conditions are mainly small animals: apybaras, armadillos, Brazilian skunks, wild pigs, tapirs and sloths.

Many species of birds in the area came originally from other regions. Two are endangered species, Nabu and Juriti. Reptiles and insects occur on the transition area.

Hydrography
Both projects are localised within the Gurupi River Basin. The major rivers are Guajupara and Piquia which are extensively used by the population of Pequia village for basic needs. The Pequia rivers are going through an accelerated eutrophication process. The river is covered by assai palm trees that putrify the water.

In hydrological terms, the flow of Piquia river was calculated to be 20 000 m^3/h. The river width is 12–15 m, with a maximum depth of approximately 2 m.

8.19 SOCIAL AND ECONOMIC SYSTEMS

8.19.1 Agricultural Development
Two groups of land for agriculture development occur in the area. The first has good conditions for specific crops such as sugar cane, manioc, sweet potato, beans, cotton, soyabeans and latex. The second group is classified for cattle farming and sylviculture. The Federal Government has implemented the Alcohol Programme in the region, towards the extensive plantation of sugar cane, broomcorn (sorghum) and manioc. The programme seeks the production of alcohol fuel for motor cars.

8.19.2 Social and Economic Data of Pequia Village
Pequia village is a small settlement adjacent to the BR-222 motorway. It lies within the delimited influence area of both plants. It is involved directly with the plants because of its proximity to the Carajas railway.

The imigrant workforce for the operation of the plant, however, will be recruited from Pequia which has a population of 2000 living in 400 houses which have no basic sanitation or water supply. Only a few houses have electricity. Little by little, the population are encroaching on the BR-222 motorway security area by constructing houses.

The growing health problems in the village are a source of worry to the State Public Health Secretariat. For this reason, a medical station to assist the local population has been built in the area. The major health problems are generated from water contamination by household wastes and venereal diseases, due to the incidence of prostitution.

The population of Pequia depends upon the river Piquia for drinking water, cooking, washing and bathing including recreational activities and fishing. The Carajas railway and the BR − 222 motorway are the main transport routes for the population, which links main cities, towns and villages in the region. The main economic activities in the Pequia and the surrounding areas are basically the commerce of milk and derived products, small shops, woodmills and odd jobs.

8.20 IDENTIFICATION AND ASSESSMENT OF ENVIRONMENTAL IMPACTS

The construction and operation phases of the industrial plants are the basic activities considered for generating environmental impacts on the natural, social and economic systems. These impacts were identified and assessed in a methodological approach.

8.20.1 Methodological Approach
The matrix method was used for environmental evaluation. It was structured in order to relate the project's basic actions (1, horizontal axis) with the environmental factors listed (2, vertical axis), and organised in the three categories natural, social and economic systems. The interaction between (1) and (2) represents an impact (see Fig. 8.11).

8.20.2 Classification of Impacts
The environmental impacts were classified in 10 different levels shown in Table 8.5.

8.20.3 Importance of Impact
The degree of importance was established in two different levels:

- very important negative impact
- important negative impact
- very important positive impact
- important positive impact

8.20.4 List of Environmental Impacts Identified
Positive impacts
- social development
- direct and indirect employment
- specialization of man-power
- increase product circulation
- supply diversified industries
- national and international markets

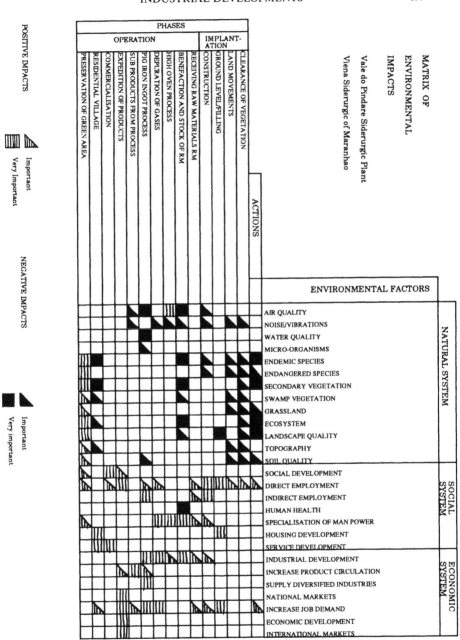

Fig. 8.11. Matrix of environmental impacts.

Table 8.5
Classification of impacts

Positive	(+)	Reversible	(r)	Local	(L)
Negative	(−)	Irreversible	(Ir)	Regional	(R)
Long-term	(Lt)	Direct	(D)		
Short-term	(St)	Indirect	(Id)		

- housing development
- service development
- industrial development

- increase employment demand
- preservation of green areas
- development of landscape projects

Negative impacts

- degradation of air quality
- increase of noise and vibrations
- degradation of water quality
- degradation of microorganisms
- degradation of endemic and endangered species

- degradation of secondary and swamp vegetation
- degradation of grassland
- degradation of ecosystems
- degradation of topography and landscape quality
- degradation of soil quality

8.21 COMMENTS ON THE MOST IMPORTANT ENVIRONMENTAL IMPACTS

8.21.1 Positive Impacts

Increase job demand

The development of the industrial plants will generate a large amount of direct and indirect employment for regional and mainly local population. Most of the jobs will be given to the population of Pequia village. This will affect positively the quality of life for the population involved with the plants. Classification of impact: Lt, Ir, D and Id, L and R.

Specialisation of man-power

Both enterprises will develop training programmes for specialist man-power that will work in the operation phase. The training will be basically on maintenance, operation processes and monitoring. The training also includes fiscalisation of preserved green areas. Classification of impact: Lt, Ir, D, L.

Development of Residential Villages

The residential villages will be constructed for the workers and their dependents. These settlements will have basic urban infrastructure and basic services including a recreational area for sports and leisure. Classification of impact: Lt, Ir, D, L.

Industrial development
Industrial development in the area is one of the major goals of the Great Carajas Project (PGC), developed by the Federal Government. On the other hand, the location of both projects are part of the industrial district of Acailandia, designed by the planning sector of the State Government.

Industrial development in the area is also part of an integrated program between the Carajas mining project and the Carajas railway. Both projects were established by Vale do Rio Doce Company (CVRD). Classification of impact: Lt, Ir, D and Id, L and R.

Increase product circulation
The expansion of production of pig-iron has a direct relationship with the demand for products on the national and international markets. This fact will increase circulation of products in order to supply steel plants and foundries. Classification: Lt, R, D and Id, L and R.

Preservation of green areas and development of landscape value
One of the major programmes regarding social–environmental issues is the development of ambient landscape design and the preservation of representative green areas. That is, the secondary vegetation, assai palm trees and representative species of local trees. The programme includes afforestation of adjacent areas for sustained charcoal exploitation. Classification: Lt, R, D, L.

8.21.2 Negative Impacts
Air quality
The risk of air pollution will occur if the filtering system for gases from high ovens and gledons does not function efficiently. In this case and depending on the wind intensity and direction, the surrounding areas will most probably be affected, first, the assai palm trees area, the secondary vegetation, grassland and part of Pequia village. The greatest risk of contamination will fall upon the workers of the plants and the cattle farming in the surrounding areas. Classification of impact: Lt, R, D, L.

Noise and vibration
Noise and vibration will arise from various sources: transport of raw materials by truck, screening of raw materials and the machine house. Noise from intensive trucking will be around 85 dB(A) mainly in the area of Carajas railway and Pequia village. The other source will be loading and unloading raw materials and iron ore.

Noise and vibration will come from turbo blowers (with noise around 90 dB(A) and the screening area. According to the project design, these areas will be acoustically sealed. Noise will also be limited by the landscape design project. Classification of impact: Lt, R, D, L.

Landscape design

Modifications on landscape quality will be mainly at local level for each plant. The view from the outside of the plants will be restricted to 10% of the high ovens. Classification of impact: St, R, R, L.

Human health

In terms of health impacts, the major risk will be on the plant workers and, second on the population of Pequia Village. In both cases health problems are due to high concentrations of gases (CO_2, CO, H_2 and N_2) which can cause dizziness, respiration difficulties and fainting. Respiration (lung) problems may occur from inhalation of dust (particulated material) from unloading and screening raw materials.

The sources of air pollution are the high ovens and gledons. When any problem or accident occurs in the plants, the risk of air pollution will be greater if the northwest wind is blowing.

Noise and vibrations can affect human hearing if basic equipment is not used in the machine house. Special equipment should also be used during unloading and screening of raw materials.

Human health may also be affected if solid wastes (fine ore and fine charcoal) are drained to the Piquia river by torrential rains. This will degradate water quality, ichthyofauna, and affect the population of Pequia Village downstream. Classification of impact: St, R and Ir, D, L.

Vegetation/endemic species/ecosystems

Fauna, flora and ecosystems are so related that when one is affected the other two also suffer. The selected area (and surroundings) for the implantation of industrial plants has suffered considerably in the early days from intensive hunting, and later on, timber and charcoal exploitation. This has caused forced migration of endemic fauna to adjacent regions.

The implantation activities, clearance of vegetation and land movements, of the plants will not cause great impacts on the ecological components. But the risk exists on the operation phase by air pollution (if the depuration system has problems). On the other hand, soil quality may also be affected if solid wastes are carried by torrential rains. Classification of impact: St, R, D and Id, L.

Water quality

The water quality of Piquia river can be affected in three possible ways: first, if the refrigeration water from the high ovens is returned to the river with a higher temperature; second, if chemical components are added to refrigeration water; and third, if the solid wastes (fine charcoal, fine ore and slag) are carried to the water source (including groundwater) by torrential rains. Classification of impact: St, Ir, D and Id, L.

8.22 ENVIRONMENTAL PROGRAMMES

Special environmental programs should be developed to mitigate and monitor adverse impacts, and to develop beneficial ones.

8.22.1 Mitigation Measures

The industrial plants were designed to have an efficient industrial production without affecting the surrounding environment. This includes the installation of special equipment to avoid major risk on environmental quality standards.

Regarding air quality, both plants will have emission control units to prevent concentrations above legal standards. Most of the gases will be recycled by the gledons for re-heating the high ovens.

The noise pollution from aerodynamic and hydrodynamic sources (machine house and other sectors) will be controlled by acoustic seals. The external noise will be compensated by the plantation of a belt of trees around both plants.

The landscape quality will be improved by afforestation of endemic species and seed grass on uncovered slopes. Certain representative areas will be preserved to maintain ecological diversity.

8.22.2 Monitoring Programmes

Air quality

Two types of monitoring air quality should be developed: (a) emission source testing and (b) atmospheric monitoring. In both cases the location of monitoring devices, the type of equipment, the duration of sampling, and pollutant discrimination are of paramount importance in quantitatively appraising air quality.

Water quality

Flows should be measured to determine the quantity of the waste water being discharged. The location of a sampling station should be selected, assuring the collection of a well-mixed representative sample. Periodicity and qualitative analysis are also important in water sampling.

Solid wastes

The disposal of solid wastes should be monitored in terms of quantitative measurements of the solid wastes handled: description of solid waste material received; major operational problems; dust and control efforts; and quantitative and qualitative effectiveness of gas and leachate control; gas sampling and analysis, ground and surface water quality sampling and analysis upstream and downstream of site.

Human health

Periodical medical examinations should be applied to workers for biological and physical monitoring, and in particular, for those employees working in areas with excess heat, CO_2, noise and relevant concentrations of particulated material. The employees should be trained to use special equipment for accident prevention, first aid, in fire accidents and personal security. The local population should also be periodically examined regarding water and air contamination.

8.23 DEVELOPMENT OF SOCIAL/ENVIRONMENTAL IMPACTS

Social programmes should be developed for the population directly and indirectly involved with the industrial plants, in relation to the following:

- development of basic educational schemes with the State Government, for population involved with the plants;
- employment of local workforce during enlargement of plants;
- development of management and maintenance plans for the villages;
- maintenance of basic services (medical), odontologic, transport and communication, for the population directly involved with the plants.

In environmental terms, specific programmes should be implemented in order to maintain environmental quality, including:

- environmental education for population involved with the plants and to put environmental issues into traditional school education;
- fiscalisation and maintenance of preserved areas and development of joint programmes with universities to study the preserved areas;
- development of a sustainable plan of charcoal exploitation between the Forest Development Institute and private project firms.

8.24 CONCLUSIONS

The base-line studies of the regional and local environment including natural resources, social and economic, in relation to the development projects were fundamental in the identification and assessment of major positive and negative impacts. This procedure allowed the following conclusions to be drawn:

- The total number of impacts identified in the matrix was 109, of which 59 were positive and 50 negative.
- There are 30 very important positive impacts against 14 negative important impacts.
- A greater number of the very important negative impacts were identified in the implementation phase (clearance of vegetation) and in the operation phase (high oven process).
- The majority of very important positive impacts were identified in the operation phase (receiving of raw materials, pig-iron ingot process, commerce of sub-products and commercialisation of pig-iron).
- Preservation of green areas also generates very important positive impacts in the natural system.
- The identification of a large number of very important positive impacts demonstrated that the development of the industrial plants will bring significant benefits for the local population and general regional development.
- The negative, important and very important, environmental impacts should be minimised by implementing environmental programs.

8.25 THE DEVELOPMENT OF AN EIS IN A DEVELOPING COUNTRY

In Brazil, as in most other developing countries, the basic conditions do not make it easy to develop an EIA system. This is due to the lack of information (base-line data), lack of specialised human resources in EIA, and the lack of experience in EIA.

The base-line (secondary) data required for an EIA in most major parts of Brazil is usually spread through many government departments, organisations and institutions. The information is difficult to obtain and most of it is on a macro-scale which is not compatible with EIA study requirements.

Experience obtained in dealing with the two pig-iron plants demonstrated the need to adapt EIA procedures and methods to the local conditions, and to improvise actions in order to obtain basic data for an environmental evaluation.

The main constraints in developing the EIS are:

- lack of knowledge regarding the political and technical requirements of the State Environmental Agency (SERNAT) EIA procedures;
- to obtain base-line data for developing an EIA;
- to obtain information concerning specific EIA topics;
- the short period of time given by the developer for preparing the EIS.

Practical solutions for each of these problems were worked out. First meetings were arranged with SERNAT to discuss the best way to develop an EIA. The second step was to research base-line data for both projects and the surrounding environment. This was developed in three steps:

- involving gathering the maximum information on both projects in viability studies, layouts, charts, maps and socio-economic data;
- researching secondary data regarding the projects' surrounding environment, local and regional, from the government and private organisations and institutions;
- obtaining first-hand experience by visiting the project sites; this involved planning site visits in conjunction with field activities. The main field activities were:
  - strategic photographic record of location of plants and adjacent areas;
  - checklist and description of environmental (natural, social and economic) factors at each point that photographs were taken;
  - interviews with local population concerning their life-style (socio-economic values) and their relationship with the surrounding environment; and
  - development of sketches of relevant environmental factors observed, using location maps of plants or other sources.

After the field work, all information researched was organised to comply with the EIA guidelines from the Federal Environmental Agency. A specific questionnaire was prepared for each member of the team (biologist, geographer, sociologist and industrial engineer) in order to evaluate the potential impacts from the various specialised field studies.

The short period of time allowed for the study was partially mitigated by developing an integrated work programme with SERNAT, and by simplifying the study as a whole, by focussing mainly on the most relevant topics.

REFERENCES AND BIBLIOGRAPHY

Braun, R.A. (1987) *Relatorio de Impacto Ambiental — RIMA*, Companhia Siderurgica Vale do Pindare, Rio de Janeiro.
Braun, R.A. (1987) *Relatorio de Impacto Ambiental — RIMA*, Viena Siderurgica do Maranhao SA, Rio de Janeiro.
CONAMA (1986) *Resolucao No 001 de 23 de Janeiro de* 1986, Brasilia, DF.
IBGE (1984) *Atlas do Maranhao*, Rio de Janeiro.
PGC (1981) Programa Grande Carajas, *Aspectos Fisicos, Demograficos e Fundiarios*, Rio de Janeiro.
The World Bank (1984) *Environmental Guidelines*, Washington, DC.
The World Bank (1984) *Occupational Health and Safety Guidelines*, Washington, DC.

PART V: Industrial Development: The Seveso Accident

8.26 INTRODUCTION

An uncontrollable exothermic process took place on July 10, 1976, during the synthesis of trichlorophenol at the Givaudan-La Roche ICMESA plant at Seveso, 30 km north of Milan. The temperature and the pressure rose inside the reactor causing a safety device to blow out. At the moment of the environmental release, the principal volatile compound inside the reactor was ethylene glycol. Small particles may have been carried out by the glycol vapour during its initial boiling phase. It has been estimated that nearly 3000 kg of organic matter left the reactor, including at least 600 kg of sodium trichlorophenate with some amount of 2,3,7,8-tetrachloro-dibenzo-p—dioxin (TCDD) (Inquiry, 1979; Di Domenico et al., 1982; Pocchiari et al., 1983). A toxic cloud was thereby released into the atmosphere, which rose to some 50 m, then subsided and fell back to earth, contaminating about 1800 ha of a densely populated area, called the Brianza of Seveso (Fig. 8.12). Leaves of plants near the ICMESA plant, courtyard animals and birds were seriously affected, many dying within a few days after the accident. At the same time, dermal lesions began to appear among the inhabitants of the area. Only 9 days after the accident, it was assessed that TCDD was present in various types of environmental samples collected near the ICMESA plant. On 26 July, 1976, the Italian Authorities evacuated 179 people from a 15-ha area immediately southeast of the plant. A few days later, further sampling of soil and vegetation indicated the presence of TCDD in a more extended area. As a

208

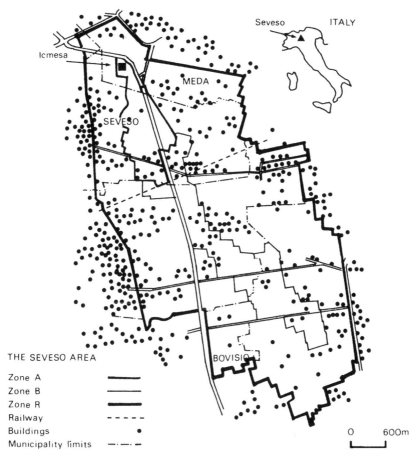

Fig. 8.12. The Seveso area Zones A, B and R at their maximum extension, showing major built-up areas (●) and surrounding farm lands. The ICMESA plant appears within Meda municipality boundaries near the Meda-Seveso borderline.

consequence, all the inhabitants (733 people) in a wide area, coded Zone A (approximately 100 ha) were evacuated (Fig. 8.12). Moreover, inhabitants of the surroundings (Zones B and R) were subjected to a number of hygiene regulations, including the prohibition to farm and consume local agricultural products and keep poultry and other animals. Zone B (270 ha) was the natural extension of Zone A and exhibited lower TCDD contents (Fig. 8.12). Both Zones A and B were enclosed by Zone R (1430 ha), exhibiting undetectable or near-detection TCDD levels. The Zone A–Zone

B and the Zone B–Zone R borderlines approximately ran along the 50 $\mu g/m^2$ and 5 $\mu g/m^2$ TCDD mean concentration lines respectively. The Zone R boundaries were set at undetectable TCDD levels (formally, $< 0.75 \mu g/m^2$.

The emergency phase lasted up to the end of the first week of August. After this period, a more precise and detailed assessment of the extent and environmental effects of TCDD environmental distribution, as well as of TCDD environmental fate and persistence, was planned.

8.27 TCDD ENVIRONMENTAL DISTRIBUTION MAPPING
8.27.1 Zone A
The standard soil sampling was carried out by coring the soil surface with a cylindrical (7 cm diameter) sampler vertically introduced into the ground to extract a core specimen (7 cm high).

The first systematic sampling of Zone A was carried out on 11–13 August, 1976; 108 sampling points were examined. A limited area close to the ICMESA plant, exhibited the highest TCDD values. TCDD levels detected in Zone A ranged from the analytical threshold to about 20×10^3 $\mu g/m^2$. A subsequent more extensive mapping of Zone A was completed in December 1976, based on 405 sampling points.

A further mapping was carried out on March 1978, based on replicated sampling.

8.27.2 Zones B and R
Zones B and R were extensively monitored with sampling campaigns carried out during the first year after the accident. Further samplings were carried out in the successive years.

8.27.3 Measurement Reproducibility of TCDD Soil Level
The replication of sampling and chemical determinations allowed the estimation of measurement reproducibility of TCDD soil levels.

The statistical distribution of TCDD soil concentrations was assessed to be consistent with a 'log-normal' statistical distribution (the logarithms of TCDD levels appeared to be 'normally' distributed).

The standard deviation of replicated log-transformed data was assessed to be in the order of 0.7–1 natural logarithms, corresponding to a factor of about 2–2.7 for original data (Di Domenico et al., 1980, 1982).

8.27.4 Vertical Distribution of TCDD in Soil

Investigations on the TCDD vertical distribution in soil were carried out mainly in Zone A, where higher TCDD concentrations allowed detectable recoveries at greater depths. Results from 1976 and 1977 surveys indicated steep vertical gradients which varied significantly only during the first 5–10 months after the accident (Domenico et al., 1980; Pocchiari et al., 1983).

8.27.5 TCDD Levels in Ground Water

The monthly determinations on ground and pipeline waters, carried out since August 1976, have consistently yielded negative results, even with an analytical threshold of 1 pg/1. Levels in the order of the ppt were found in sediments of the water course passing through the most contaminated area (Pocchiari et al., 1983).

8.27.6 TCDD Levels in Atmospheric Particles

'High-volume' samplers (particles up to 100 μm) and 'dust-fall jars' (particles larger than 10–15 μm) were used to monitor the air-borne dust TCDD content. The TCDD concentration in the air-borne dust was assessed to be in the order of 10–100% of the one measured in the corresponding upper soil layer (0–7 cm deep). The comparison of TCDD concentration in dust and soil indicated that air-transported TCDD did not significantly affect the TCDD soil level.

Data from dust-fall jars indicated a decreasing TCDD concentration in dust with increasing distance from the most contaminated area of Zone A.

Seasonal variations of dust deposits were detected, with a maximum at the beginning of summer.

The maximum assessed TCDD level in inhalable dust, detected in a site close to the most contaminated part of Zone A immediately after the environmental contamination, was about 0.06 pg $TCDD/m^3$ of air (Di Domenico et al., 1980).

8.27.7 TCDD Levels in Plant Tissues

The first systematic surveys of TCDD content of plants indicated that the aerial parts of vegetation directly contaminated by the TCDD-containing cloud were particularly affected, while the underground parts exhibited TCDD levels comparable with the ones detected in the corresponding soil (Cocucci et al., 1979).

A further study confirmed the ability of plants to absorb and translocate TCDD (Cocucci, 1980). In the case of plants not directly contaminated by the toxic cloud and grown on TCDD-containing soil, available data suggest that the TCDD concentration in hypogeal tissues is about one order of magnitude higher than the one in epigeal parts. In particular, an average level of 0.01–0.02 ppb was found in the epigeal part of fodder growing in an experimental field in Zone A whose TCDD average level in soil was about 2–3 ppb. The TCDD average level in the hypogeal part of this fodder was about 6-fold higher than in aerial parts.

The ratio between TCDD level in plants and TCDD level in soil was about 0.03 for fodder aerial parts and about 0.6 for bulbous plants hypogeal parts (Zapponi et al., 1986).

Experiments effected outside of the Seveso area, using plants cultivated in greenhouses, indicated a ratio TCDD in plant/TCDD in soil higher in low level contaminated soils than in high level contaminated soils (Facchetti et al., 1984, 1986).

The extremely low TCDD solubility in water, recently indicated to be of the order of 8 ppt (Adams and Blaine, 1986), can be the key to explain this finding. In the case of high TCDD concentrations in soil, the soil-contained water may be expected to be saturated by TCDD; this implies also a saturation of the TCDD transport process by water from soil to plant roots.

8.27.8 TCDD Environmental Persistence
The comparison of TCDD levels detected in 44 sampling sites of Zone A at 1, 5 and 17 months after the ICMESA accident, indicated a statistically significant and sharp decrease of TCDD soil content during the first 6 months after the accident. No further decrease was detected after this period. The TCDD permeation below the first 1–2 cm deep soil layer may explain these findings; the solar ultraviolet rays could no longer act on and therefore degrade the compound.

8.27.9 The Physical/Chemical Properties of TCDD and its Environmental Behaviour
TCDD has a strong tendency to be absorbed to soil organic carbon (its sorption constant is in the order of 500 000), an extremely low solubility in water (possibly 8 ng/l; Adams and Blaine, 1986) and a very low vapour pressure at environmental temperatures.

At the equilibrium state, the environmental partition models indicate that more than 99.9% of the environmentally released TCDD may be expected to be absorbed to soil organic carbon. This may easily explain very limited or absent vertical mobility of TCDD after the first transitory period.

8.27.10 The Health Impact of the TCDD Accidental Release in the Seveso Territory

The epidemiological study of the health consequences of the Seveso accident was difficult, due to the heterogeneity of the exposed population and of the exposure pattern.

The mortality studies in the 1975–80 period do not indicate remarkable differences between zones at different pollution levels. The mortality pattern was typical of industrial areas, with cardiovascular diseases and cancer as principal causes of death.

The data available indicate some evidence of an increase in absorptivity rates in polluted areas, from 1976 to 1978, possibly attributable to the TCDD environmental contamination (in 1978: Zones $A + B + R =$ 12.75%; external area = 9.8%; $p < 0.05$) (Bruzzi, 1983).

Higher rates of some groups of malformations were observed in polluted zones, although not statistically significant.

No definitive conclusions are at present available relative to cancer incidence, possibly attributable to the accident, due to the limited amount of time elapsed since the accident.

Several studies have indicated effects on liver, blood lipids and on the peripheral nervous system during the first period after the accident.

A detailed evaluation of the health effects caused by the Seveso accident is not possible here. Although the Seveso accident was not the feared catastrophic event, 'it is fallacious to state that the ICMESA accident passed over the Seveso population with no effect other than the occurrence of chloracne' (Bruzzi, 1983).

8.28 CONCLUSIONS

The above data show that the TCDD impact is mainly exerted on the soil compartment, as well as on plants and animals in direct contact with it. The data relative to the vertical mobility of TCDD in soil indicate that an initial significant permeation took place shortly after the accident, as a consequence of the first heavy rains. After this first 'transitory' process no substantial variations were detected.

As far as the TCDD environmental persistence is concerned, a significant decrease of TCDD soil levels was detected only in the first six months after the accident.

The absence of TCDD environmental degradation after this period can be ascribed to the limited availability of this compound to the action of solar ultraviolet rays.

The considerations indicate that the initial period was particularly important in determining the environmental fate of the accidentally released TCDD and the consequent human exposure pattern in the affected area.

The health effects of the Seveso accident were generally not so serious as initially expected; this was in large part due to the prudential measures adopted. However, this does not mean that the accident was without health consequences.

The future will give further data.

REFERENCES AND BIBLIOGRAPHY

Adams, J.A. and Blaine, K.M. 1986 A water solubility determination of 2,3,7,8 – TCDD, *Chemosphere*, 15(9–12), 1397.

Cocucci, S., Di Gerolamo, F., Verderio, A., Cavallaro, A., Colli, G., Gorni, A., Invernizzi, G. and Luciani, L. (1979) Absorption and translocation of tetrachlorordibenzo-*p*–dioxins by plants from polluted soil, *Experientia* 35(4), 482.

Cocucci, S. (1980) Absorption, translocation and elimination of TCDD by plants in contaminated soil, International Steering Committee 3rd Meeting, March 30–April 1, Segrate (Milano) Italy.

Di Domenico, A., Silano, V., Viviano, G. and Zapponi, G.A. (1980) Accidental release of 2,3,7,8 – TCDD environmental contamination at Seveso. In: O Hutzinger et al. (eds), *Chlorinated Dioxins and Related Compounds. Impact on the Environment*, Pergamon Press, Oxford, pp. 47–54.

Esposito, M.P., Tiernan, T.O. and Dryden, F.E. (1980) Dioxin. EPA – 600/ 2.80.197, Cincinnati, USA.

Facchetti, S., Balasso, A., Fichtner, C., Frare, G., Leoni, A. and Mauri, G. (1984) Studi relativi all assunzione di TCDD da parte di alcune specie vegetali, Atti del Convegno La Risposta Tecnologica Agli Inquinamenti Chimici 20–21–22 settembre 1984, Milano (Italia), 231.

Facchetti, S., Balasso, A., Fichtner, C., Frare, G., Leoni, A., Mauri, C. and Vasconi, M. (1986) Studies on the absorption of TCDD by some plant species, *Chemosphere*, 15(9–12), 1387.

Hay, A. (1982) *The Chemical Scyte*, Plenum Press, New York.

McCall, P.J., Laskowski, I.A., Swann, R.L. and Dishburger, H.J., Estimation of environmental partitioning of organic chemicals in model ecosystems, *Residue Rev.*, 12, 231.

McKay, D. and Paterson, S. (1982) Fugacity revisited, *Environ. Sci. Technol.*, 16(12), 654.

Pocchiari, F., Di Domenico, A., Silvano, V. and Zapponi, G.A. (1983) Environmental impact of the accidental release of tetrachlorodibenzo-dioxin (TCDD) at Seveso (Italy). In: Coulston and Pocchiari (eds), *Accidental Exposure to Dioxin Human Health Aspects*, Academic Press, New York, pp. 5–35.

Pocchiari, F., Cattabeni, F., Della Porta, G., Fortunati, U., Silano, V. and Zapponi, G.A. (1986) Assessment of exposure to 2,3,7,8-tetrachlorodibenzo-*p*–dioxin (TCDD) in the Seveso area. *Chemosphere*, **15**(9–12), 1851.

PART VI: Health Impacts of Water Development in Turkey

8.29 INTRODUCTION

Turkey is rich in water resources and is surrounded by sea on three sides with a coastline of over 8300 km. The land surface is approximately 800 000 km² and is watered by innumerable rivers varying in length from 30 to 2800 km. There are 48 natural and 29 artificial lakes greater than 5 km².

Turkey is mainly a mountainous country with an average altitude of 1030 m. The Anatolian peninsula is bordered by high mountain ranges in the north and south. Between these ranges and the sea there are narrow coastal plains while the inland is divided into a number of plateaux. The Cukurova located in the south is one of them.

8.30 LOWER SEYHAN IRRIGATION PROJECT AND MALARIA PROBLEM IN TURKEY

One of the greatest water-related projects in Turkey was started in the Cukurova in the early 1950s; it is known as the Lower Seyhan Irrigation Project. Following the development of the irrigation project, malaria became a health problem.

8.30.1 Geographical Description of the Area
The Cukurova is located in the southern part of Turkey, along the Mediterranean coastal belt, and is watered by three main rivers running north to south, the Berdan, Seyhan and Ceyhan.

The Lower Seyhan Plain is limited by the Ceyhan and Berdan Rivers in the east and west, respectively. Together with the Seyhan river, these rivers form a flat alluvial delta.

The flat landscape of the plain is interrupted by small and isolated hills. A narrow band of sand dunes runs parallel to the coastline and obstructs the natural drainage of the rainwater into the sea. The dunes are interspaced with lagoons with swampy and poorly drained areas. All excess water from the irrigation system is collected by three main drainage canals reaching the sea or two lagoons named Akyatan and Tuz G818. The length of Akyatan lagoon is around 16 km and it covers about 5000 ha.

8.30.2 Climatological Features of the Area

The plain has a mediterranean climate with hot summers and moderate winters. The humidity is high throughout the year. The highest precipitations occur in December and these slowly decrease until August which is the driest and warmest month of the year. Rainfalls start in September again and increase, reaching their maximum in December or sometimes in January.

8.30.3 Historical Background of Malaria in Turkey and in Adana

Malaria had been one of the major health problems for centuries in Turkey, although numerical information before 1925 is not available.

After the foundation of the Republic of Turkey, the highest priority has been given to malaria as the most important health problem of the community. A scientific group was established in 1924 to prepare a report on malaria including outlines of an anti-malaria programme for the country.

After discussions on this report in the First National Medical Congress held in Ankara in 1925, necessary measures to fight against malaria were taken by the government and ultimately two laws related to malaria were promulgated in April 1926. Compulsory participation in a three-month training course in Malaria Institutes for newly graduated medical doctors was established by Law No 826 while Law No 839 was aimed directly at the implementation of actions against malaria. After the promulgation of these laws, the Adana Malarian Institute was established as a training and research centre in 1926. The number of malaria cases reported from the whole country between 1925 and 1946 is shown in Table 8.6.

The main activities against malaria during that period were the following:

(a) elimination of marshes by drying, drainage or re-forestation;

Table 8.6

Malaria in Turkey during the period 1925–46

Year	Population in census years	No. of malaria cases	API (%000)	Year	Population in census years	No. of malaria cases	API (%000)
1925		1434		1937		69850	
1926	(x)	14791		1938		81702	
1927	13648270	10190	74·7	1939	(xx)	120060	
1928		9928		1940	17820950	115683	649·1
1929		36186		1941		94534	
1930		45635		1942		146077	
1931		61241		1943		115546	
1932		72500		1944		80387	
1933		50609		1945	18790174	16739	89.1
1934		48744		1946	(xxx)	16373	
1935	16158018	40842	252·8				
1936		62466					

(x) Beginning of antimalaria programme.

(xx) Beginning of Second World War (shortage of quinine, large population movement).

(xxx) Establishment of the Directorate General of National Malaria Programme.

(b) applications of larvicides to water collections (Paris green or fuel oil);

(c) free distribution of quinine to the population under risk and treatment of diagnosed malaria cases.

In 1946 the Directorate General of National Malaria Programme was established within the Ministry of Health and Social Assistance. Until that date, anti-malaria activities were directed by a special division of the Directorate General of Public Health.

The malaria situation during that period is given in Table 8.7.

The national malaria eradication campaign was initiated in 1957 in Turkey. The results obtained from that campaign are shown in Table 8.8.

In spite of the excellent results obtained from the eradication campaign, malaria unfortunately recurred and reached epidemic proportions in a very short time after the campaign, as shown in Table 8.9.

The number of cases reported from Adana was 5128 in 1958 (45.73% of total malaria cases). Residual indoor spraying was the main tool of the eradication programme; this was carried out using dieldrin and DDT.

The attack phase took place in 1963 and the area then moved to the consolidation phase. Only 25 malaria cases were reported from Adana

Table 8.7
The malaria situation during the period 1947–56

Year	Population in census years	No. of malaria cases	API (%000)	Year	Population in census years	No. of malaria cases	API (%000)
1947		5979		1952		8400	
1948		7298		1953		5227	
1949		4973		1954		2489	
1950	20947188	4211	20·1	1955	24064763	1494	6·2
1951		20132		1956		1573	

Table 8.8
Malaria in Turkey during the period 1957–70

Year	Population in census years	No. of malaria cases	API (%000)	Year	Population in census years	No. of malaria cases	API (%000)
1957		5536		1964		5081	
1958		11213		1965	31391421	4587	14·6
1959		7305		1966		3793	
1960	27754820	3092	11·1	1967		3975	
1961		3498		1968		3318	
1962		3594		1969		2173	
1963	(x)	4365		1970	35605176	1263	3·5

(x) End of the attack phase.

between 1961 and 1963. The proportion of malaria cases reported from Adana has been 47.13% in 1983 and 53.87% in 1983 in total.

The stages of activities in the development project of the Lower Seyhan Plain can be listed as follows:

(1) the construction of a dam on the Seyhan river to store the water for hydro-electric and agricultural purposes;
(2) the establishment of a spillway for excess water;
(3) construction of belt or primary irrigation canals in order to distribute the water throughout the plain;
(4) construction of secondary and tertiary irrigation canals for irrigation of fields;
(5) construction of primary, secondary and tertiary drainage canals for draining excess water from the fields.

Table 8.9
The malaria situation in Turkey after the eradication campaign

Year	Population in census years	No. of malaria cases	API (%000)	Year	Population in census years	No. of malaria cases	API (%000)
1971		2046		1978		87867	
1972		2898		1979		29324	
1973		2438		1980	44736957	34154	80·0
1974		2877		1981		54415	
1975	39147855	9828	20·0	1982		62038	
1976		37320		1983		66681	
1977		115512		1984		45733	

Table 8.10
Design features of irrigation canals (trapezoidal shape)

Canal type	Length (km)	Flow rate (m³/s)	Width (m) Top	Width (m) Bottom
Primary		5·0–10·0	15–25	2·0–9·0
Secondary		0·5–5·0	8–15	0·8–2·0
Tertiary	1·537	0·05–0·5	1·0–1·5	0·4–0·8
	3·553			

Table 8.11
Design features of drainage canals (trapezoidal shape)

Canal type	Length (km)	Flow rate (m³/s)	Width (m) Top	Width (m) Bottom
Primary	300	Variable	15–30	5–25
Secondary	300	Variable	10–15	2–5
Tertiary	710	Variable	5–10	1–2
	1310			

The numerical data about irrigation and drainage systems are given in Tables 8.10, 8.11 and 8.12.

There are two national institutions in charge of the design, construction, operation and maintenance of irrigation and energy projects in Turkey:

Table 8.12
Volume of water used for irrigation (in thousand m³)

Area	May	Jun	Jul	Aug	Sep	Oct	Total
Left bank	26381	84453	146553	171656	92171	27217	547431
Right bank	42329	40409	101860	105167	63333	26335	379433
Total	68710	124862	248413	276823	155504	53552	926864

DSI (State Hydraulic Works) and Topraksu (Directorate General of Soil and Water). DSI is responsible for water reservoirs, irrigation and main drainage canals and energy plants as well as for cities with more than 100000 population. Topraksu is in charge of the construction of drainage canals in fields and of providing technical assistance to farmers for the implementation and proper agricultural procedures in order to obtain the maximum benefit per unit of cultivated area.

The results of activities can be summarised as follows:

(1)   population movements from localities to be covered by water and their resettlement around newly irrigated areas;
(2)   increase of the productivity of irrigated lands;
(3)   introduction of insects and different kinds of insecticides into the area and creation of resistance of vectors to insecticides;
(4)   increasing needs for manpower to prepare plantable fields expanded by irrigation to harvest the crop that increased after the implementation of the project;
(5)   population movements from poorer parts of the country towards newly developing area as seasonal workers, most of whom come from areas where unnoticed malaria epidemics still occur;
(6)   settlement of seasonal workers along the canals were due to slopes less than 1% and due to the vegetation, water collections have become efficient breeding places for malaria vectors;
(7)   introduction of malaria parasites to the local vector *An. sacharovi* which has a great capacity for transmitting the disease.
(8)   initiation of industries to work on local products increased by the agricultural development;
(9)   increasing needs for labour force to work in industries;
(10)   more population movements towards industrial activities, resulting in the increase of the population of Adana province (from 1240475 in

1975 to 1487743 in 1980 with an annual increase of 36.08 per thousand).

(11) establishment of unhealthy settlements around towns for the newcomers;

(12) establishment of new, high apartment buildings in order to meet the housing needs of the newcomers to the area and increasing number of breeding places in underground floors of these buildings, due to the high level of the water table and to deep basement excavations;

(13) transmission of malaria parasites to non-immune local people, and finally, resurgence of malaria in a developed area with the aid of other wrong administrative and operational activities.

This is a short summary of the story of malaria epidemics in Adana and in the Lower Seyhan Plain of Turkey.

8.31 RESULTS

In the mid-1960s, the arrival of malaria infected workers from the southeastern part of Turkey where malaria transmission still occurred, to the Cukurova, was not thought of as a danger for creating a new epidemic in the area where conditions were most favourable to it.

Due to the very low number of malaria cases reported from the entire country, national and international authorities were led to believe that the disease was totally under control.

In 1969 the organisational structure of the malaria programme became weakened by a reduction not only in financial support, but also in the manpower. The fact that during 1970 the number of cases reported from the Cukurova had increased from 49 to 149 passed unnoticed.

In conclusion, malaria is an example of the dimensions of health impacts of water development projects in Turkey. The example can similarly be extended to water-borne, water-based, water-washed and other water related diseases in the country.

Annex A

Selection of Projects for EIA (Screening)

In some countries, no guidance exists to determine whether an EIA is required: a decision, in these cases, often being dependent upon the scale of the proposal, its environmental setting and the likely degree of public opposition. In other countries, various screening procedures exist and the differing requirements of each procedure has resulted in considerable disparities as to the number of EIAs undertaken in individual countries.

In this section, some general considerations relating to screening are presented before attention is focused on a number of methods which may be used to determine when an EIA is required.

General Considerations
During the development of screening procedures the following objectives require consideration:

(i)   clear identification of projects requiring an EIA;
(ii) quickly and easily used to avoid unnecessary delay. Three project categories may be identified as a result of screening, namely:
      (a) projects clearly requiring an EIA;
      (b) projects not requiring an EIA; and
      (c) projects for which the need for an EIA is unclear, that is, an intermediate category.

One solution to the uncertainty regarding the need for an EIA for 'intermediate category' projects is to subject all such projects to an EIA even though it might be shown that no significant environmental effects were likely. Such an approach may prove costly in time and resources, as well as lead to criticism of the EIA procedures. An alternative is to employ a two stage screening process in which an initial screening can be used to

Table A1
Environmental impacts often associated with project size

Probable effects	Visual intrusion Direct impact on flora and fauna on the site and surroundings Lost opportunity for alternative land uses
Possible effects	Potential pollution problems Potential local employment impact Potential local transport consequences Influence on local land use patterns (induced development) Changes in social values of local communities

quickly identify those projects which do and do not require an EIA. A secondary screening can then be used subsequently to determine whether EIAs are needed for the remaining projects (see Chapter 2, Fig. 2.1).

Methods for Screening
Five main methods are available to assist screening. They are:

(i)   project thresholds;
(ii) sensitive area criteria;
(iii) positive and negative lists;
(iv) matrices; and
(v)   initial environmental evaluations.

Projects may be categorised by a number of parameters (see Table A1), the values of which will be dependent upon the project. Particular values can then be selected to determine whether an EIA might be required. For example, projects with a capital cost greater than say £5 million could then be subject to an EIA. Project parameters are, by one means or another, incorporated into the five main screening methods.

Project Thresholds
This method relies upon the establishment of thresholds for key features of the project or its environment. If a threshold were exceeded, then an EIA would be required. Such thresholds can range from environmental factors such as the amount of agricultural land used for development, to project factors such as project size, cost or infrastructure demands. Table A1 illustrates those environmental consequences which may be related to project size.

Project size may be described in the following terms:

(i)   surface area of the project;
(ii) area of land influenced by the project, e.g. by noise, zone of visual influence; and
(iii) volume and height of the project.

To apply thresholds based simply on these project factors alone could lead to anomalies, since a 1000 m² project in one area may give rise to greater adverse effects than a 5000 m² project depending on its location. It is consequently of greater utility to create general guidance for such thresholds, for example:

(i)   If the project land taken is considered to be large in comparison to the scale or character of the surrounding environment, then an EIA is required.
(ii) If the area affected is of more than local importance or if a large number of people outside the project area would be adversely affected, then an EIA is required.
(iii) If the project is highly visible, then an EIA is required.

Project cost or capital investment may be used as a threshold, since it indicates that the project may encompass the following elements:

(i)   have a large land area;
(ii) involve high value land (consequently often in densely populated or coastal areas);
(iii) have a large project size; and
(iv) require advanced machinery or processing plant.

The application of financial criteria creates the situation in which projects may cost less than the financial threshold, yet still give rise to significant impacts. Alternatively, some projects exceeding the parameter may give rise to no significant impacts. It is, consequently, important to develop appropriate thresholds which achieve the following:

(i)   identify, at the lowest financial investment level, the greatest number of projects having significant impacts;
(ii) minimise the number of projects likely to be incorrectly identified as having significant impacts; and
(iii) minimise the number of projects falling below the threshold which have significant environmental consequences.

Reliance upon one threshold may give rise to a number of incorrect decisions, it is, therefore, normal practice to link a series of different thresholds together. For example, in France a project cost of 60 million

francs is linked to size thresholds, such as housing schemes over 3000 m²
require an EIA. Other thresholds such as input–output parameters, for
example raw material requirements or pollution generation, may be
developed. Care will, however, be needed to prevent a series of thresholds
becoming cumbersome and time-consuming to use. Thresholds while,
being simple, may need frequent revision particularly in the light of
experience and inflation when financial criteria are applied.

Sensitive Areas

Since the environmental consequences of a project is a function of both the
project and the receiving environment, the sensitivity of that environment
provides a means for determining whether an EIA is required. Environmentally sensitive areas (ESAs) may be determined in two distinct, but
complementary ways. One approach is to determine the carrying capacity
of the area in relation to the degree or intensity of interference or
disturbance. This approach implies a pre-determined series of values which
describe the resilience of the environment in relation to defined perturbations. For example, the concentration of pollutants which can be
discharged without giving rise to adverse effects. A major problem with
this approach is the relationship of the various generators of pollution to
the define assimilative capacity or limit which the environment can accept.
Proposed developments giving rise to pollutant discharges below the
carrying capacity of the environment need not require an EIA. Yet, when
other polluting activities are proposed which cause the cumulative
discharges to exceed the carrying capacity, only the latter development
would require an EIA, even though its discharge may be of a lower amount
than the former development. This approach requires considerable
amounts of information concerning the environment and, given mans'
limited understanding of many environments, criteria indicating a carrying
capacity in relation to a particular pollutant or development activity is
likely to be subject to error, and hence may give rise to controversy over its
application or relevance.

An alternative approach to the identification of ESAs is to determine the
importance of individual components of the area. In this approach, the
characteristics of the environment, in terms of its objective and subjective
values, rather than purely its ability to withstand perturbations, are given
emphasis. Battelle (1978) produced, for the Commission of the European
Community, a classification scheme by which ESAs could be identified (see
Fig. A1). In order to identify specific zones to be designated as an ESA,
Battelle suggested that the following characteristics should be considered:

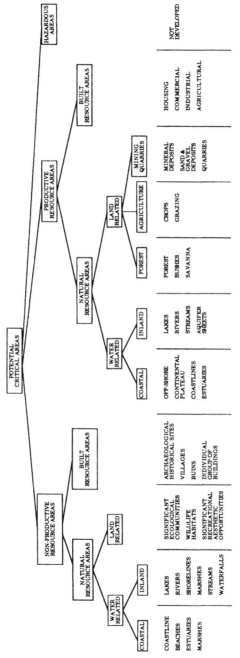

Fig. A1. List of potential critical areas (from Battelle, 1978).

Table A2
Criteria to aid identification of environmentally sensitive areas

1	Distinctive and unusual land forms
2	Importance of the ecological function of the areas to the maintenance of a natural system beyond its boundaries
3	Unusual or high quality plant and/or animal communities
4	Unusual habitat with a rarity value
5	Unusual high diversity of biological communities due to a variety of geomorphological features etc.
6	Provision of a habitat for rare or endangered species
7	Large area providing a habitat for species that require such extensive areas
8	Area location in combination with natural features providing a resource in scientific research or education terms
9	Aesthetic value of the locality

Table A3
Stages in the identification of environmentally sensitive areas (adapted from Eagles, 1981)

1	Assemble interdisciplinary study team
2	Adopt standardised criteria
3	Locate sources of information, e.g. government studies, local experts
4	Utilise sensitive area criteria to identify possible sensitive areas
5	Field validation of sensitive areas
6	Compile background files on important elements within the sensitive area, e.g. endangered species, hydrographic regime etc.
7	Final screening to identify sensitive areas
8	Delineation of sensitive areas on topographic maps
9	Detail reasons for criteria fulfilment of each delineated area
10	Publish the results

(i)	quality of the zone in terms of its own value, for example, recreational use, and in relation to other zones, for example by providing seasonal feeding grounds for wildlife from another locality;

(ii)	the abundance of the particular zone; a zone may be valued due to its uniqueness or relative scarcity due to competing demands for its use, for example a forested zone may be valued due to multiple land uses of recreation, wildlife or silviculture;

(iii)	sensitivity to change in terms of the ability of the zone to withstand or spread the consequences of human activity, for example transfer of pollutants to another zone.

Eagles (1981) also presented a series of criteria and steps to aid the

Table A4

Proposed list of projects to be subject to mandatory EIA (from
Commission of the European Community, 1980)

---

1   Extractive industry
    Extraction and briquetting of solid fuels
    Extraction of bituminous shale
    Extraction of ores containing fissionable and fertile material
    Extraction and preparation of metalliferous ores

2   Energy industry
    Coke ovens
    Petroleum refining
    Production and processing of fissionable materials
    Generation of electricity from nuclear energy
    Coal gasification plants
    Disposal facilities for radioactive waste

3   Production and preliminary processing of metals
    Iron and steel industry, excluding integrated coke ovens
    Cold rolling of steel
    Production and primary processing of non-ferrous metals and ferro-alloys

4   Manufacturing of non-metallic mineral products
    Manufacture of cement
    Manufacture of asbestos-cement products
    Manufacture of blue asbestos

5   Chemical industry
    Petrochemical complexes for the production of olefins, olefin derivatives, bulk
      monomers and polymers
    Chemical complexes for the production of organic intermediates
    Complexes for the production of basic inorganic chemicals

6   Metal manufacture
    Foundries
    Forging
    Treatment and coating of metals
    Manufacture of aeroplane and helicopter engines

7   Food industry
    Slaughter-houses
    Manufacture and refining of sugar
    Manufacture of starch and starch products

8   Processing of rubber
    Factories for the primary production of rubber
    Manufacture of rubber tyres

9   Building and civil engineering
    Construction of motorways
    Intercity railways, including high speed tracks
    Airports
    Commercial harbours
    Construction of waterways for inland navigation
    Permanent motor and motorcycle racing tracks
    Installation of surface pipelines for long distance transport

---

Table A5
Proposed list of projects which may be subject to EIA

1 Agriculture
  Projects of land reform
  Projects for cultivating natural areas and abandoned land
  Water management projects for agriculture (drainage, irrigation)
  Intensive livestock rearing units
  Major changes in management plans for important forest areas

2 Extractive industry
  Extraction of petroleum
  Extraction and purifying of natural gas
  Other deep drillings
  Extraction of minerals other than metalliferous and energy-producing
    minerals

3 Energy industry
  Research plants for the production of fissionable and fertile material
  Production and distribution of electricity, gas, steam, and hot water
    (except the production of electricity from nuclear energy)
  Storage of natural gas

4 Production and preliminary processing of metals
  Manufacture of steel tubes
  Drawing and cold forging of steel

5 Manufacture of glass fibres, glass wool and silicate wool

6 Chemical industry
  Production and treatment of intermediate products and fine chemicals
  Productions of pesticides and pharmaceutical products, paint and
    varnishes, elastomers and peroxides
  Storage facilities for petroleum, petrochemical and chemical products

7 Metal manufacture
  Stamping, pressing
  Secondary transformation treatment and coating of metals
  Boilermaking, manufacture of reservoirs, tanks and other sheet-metal
    containers
  Manufacture and assembly of motor vehicles (including road tractors)
    and manufacture of motor vehicle engines
  Manufacture of other means of transport

8 Food industry
  Manufacture of vegetable and animal oils and fats
  Processing and conserving of meat
  Manufacture of dairy products
  Brewing and malting
  Fish-meal and fish-oil factories

9 Textile, leather, wood, paper industry
  Wool washing and degreasing factories
  Tanning and dressing factories

Table A5—*contd.*

Manufacture of veneer and plywood
Manufacture of fibre board and of particle board
Manufacture of pulp, papers and board
Cellulose mills

10 Building and civil engineering
Major projects for industrial estates
Major urban projects
Major tourist installations
Construction of roads, harbours, airfields
River draining and flood relief works
Hydroelectric and irrigation dams
Impounding reservoirs
Installations for the disposal of industrial and domestic waste
Storage of scrap iron

11 Modifications to development project included in Annex 1.

identification of ESAs (see Tables A2 and A3). Regardless of the approach adopted, projects located in ESAs could be required to undertake an EIA.

The use of ESAs in a screening exercise has the advantage of being simple and easy to use, however, if it were applied alone, it does have the disadvantage of ignoring project characteristics. This may give rise to EIAs being performed on activities having no significant environmental consequences. It is, therefore, useful to balance project criteria with sensitive area criteria, for example, a project threshold of x mg/m^2 heavy metal concentration in the effluent within a fishery resource area would cause an EIA to be required, while such a concentration may be acceptable in less sensitive areas.

Positive and Negative Lists
The positive list approach can be illustrated by the proposed directive of the Commission of the European Community on EIA (Commission of the European Community, 1980). In this document two lists of projects have been proposed. The first list (Table A4) details those projects which would be subject to EIAs, while the second list (Table A5) contains those projects which shall be made subject to an EIA 'where the Member States consider that their characteristics so require'. Member States may, consequently, specify certain types of projects as being subject to an EIA, or may establish the criteria and/or thresholds necessary to determine those

projects which are to be subject either to an EIA, another form of assessment or are to be exempt from any assessment.

Positive lists may be compiled by a review of existing developments, identifying those giving rise to significant environmental damage. Some indication of the environmental importance of categories of projects may also be gained by examining the existing licensing/consent requirements. Those project categories requiring many consents may be worthy of closer examination to determine whether they should be entered on a positive list. Similarly, projects which seldom give rise to adverse environmental consequences, for example agricultural buildings, can be identified and entered on a negative list, that is a list of projects not requiring an EIA. In the case of those projects in which classification is difficult, an intermediate status can be created in which other supporting screening methods can be applied, such as thresholds or ESAs. Alternatively, all such projects for which uncertainty exists could be subjected to an EIA, although an exemption to an EIA could be granted by the appropriate authority on a case-by-case basis. In Ontario all plans, programmes, projects of activities are subject to an environmental assessment unless specifically exempted by the Minister of the Environment.

Lists are one of the simplest approaches to screening. Some research is required to prepare such lists, but they offer an easy-to-use system which is readily understood by all concerned. The preparation of lists, involves one main problem area; individual projects of the same general type of class may have considerable variations in size, plant, process and layout which may give rise to varying environmental consequences. In addition, gaining the acceptance of all parties on the entry of individual project types on particular lists (mandatory, positive, intermediate and negative) may be difficult and time-consuming.

Matrices

In Canada, the matrix has been promoted by the Federal Environmental Assessment and Review Office (FEARO) as a screening method to overcome the weaknesses of both the project and environment based screening methods, while avoiding the need for extensive studies to determine the need for an EIA. Two levels of matrices are used, the Level 1 Matrix (see Fig. A2) provides a broad screening, while the Level 2 Matrix focuses upon specific environmental impacts (see Fig. A3). During the Level 1 Matrix stage, project activities are identified which are likely to occur during the four principal development phases, namely:

Table A6
List of relevant factors for excavation

1	Extent and depth of excavation
2	Character of underlying soil (sensitive soils, permafrost etc.)
3	Proximity to noise sensitive areas
4	Disruption to traffic patterns, services, etc.
5	Requirements for water table modification
6	Surface water drainage
7	Topography
8	Quantity and type of soil
9	Nature of soil and susceptibility to water and wind erosion
10	Adjacent land use
11	Aesthetics

(i)   site investigation and preparation;
(ii) construction;
(iii) operation and maintenance; and
(iv) future and related activities.

Four main areas of potential environmental consequences are identified, namely physical/chemical, ecological, aesthetic and social aspects. Each of the interactions (potential impacts) between the 'activities' and 'effects' are then marked in the appropriate cell. For example, excavation during construction activities causes an interaction with habitats and communities (see Fig. A2).

Potential impacts identified in the Level 1 Matrix are then further defined and subdivided in Level 2 (see Fig. A3). This allows the identification of those activities which will have no effect, those for which an environmental design solution or mitigating measure can be identified, those having unknown and potential adverse effects and those having significant effects. For example the effects of excavation upon terrestrial habitats may be unknown, so a question mark is entered in that cell. On completion of this stage within the Level 2 Matrix and if significant effects are indicated, then a panel of experts is convened to review the project and coordinate an EIA. Generally, further information is needed to reduce the number of 'unknown and potential adverse effect' screening decisions made. To assist this process, background information presenting a series of 'relevant factors' is supplied for both project activities and environmental effects. Table A6 presents an illustrative list of 'relevant factors'.

In addition to the list of 'relevant factors', criteria to assist in the determination of adverse effects and indicate the type of information required were developed (see Table A7).

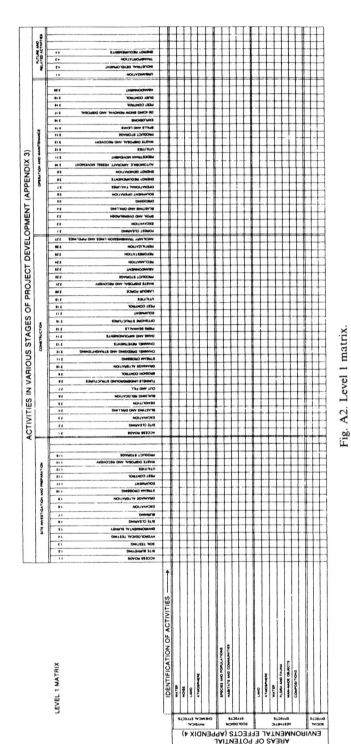

Fig. A2. Level 1 matrix.

Table A7
Criteria for making screening decisions (adapted from FEARO, 1978)

Several general criteria can be used when making a decision as to the environmental effect of an activity. These criteria are not mutually exclusive but are interrelated

Magnitude	Defined as the probable severity of each potential impact; Will the impact be irreversible? If reversible, what will be the rate of recovery or adaptability of an impact area? Will the activity preclude the use of the impact area for other purposes?
Prevalence	Defined as the extent to which the impact may eventually extend as in the cumulative effects of a number of stream crossings. Each one taken separately might represent a localised impact of small importance and magnitude, but a number of such crossings could result in a wide-spread effect. Coupled with the determination of cumulative effects is the remoteness of an effect from the source activity. The deterioration of fish production resulting from access roads could affect sport fishing in an area many miles away and for months or years after project completion.
Duration and frequency	The significance of duration and frequency can be explained as follows. Will the activity be long-term or short-term? If the activity is intermittent, will it allow for recovery during inactive periods?
Risks	Defined as the probability of serious environmental effects. The accuracy of assessing risk is dependent upon the knowledge and understanding of the activities and the potential impact areas.
Importance	Defined as the value that is attached to a specific area in its present state. For example, a local community may value a short stretch of beach for bathing or a small marsh for hunting. Alternatively, the impact area may be of a regional, provincial or even national importance.
Mitigation	Are solutions to problems available?

On the basis of these further investigations the cells marked to indicate 'unknown and potential adverse effect' are modified to record either the identification of design solutions or the realisation that significant effects will occur. If data were not available to undertake such re-classification, then an initial environmental evaluation would be required. For example, to resolve the unknown effect of excavation upon terrestrial habitat, an initial environmental evaluation would be undertaken to examine the potential effects in more detail. If at any stage, however, a decision had

LEVEL 2 MATRIX

SCREENING SYMBOLS

- ▢ NO EFFECT
- D ENVIRONMENTAL DESIGN SOLUTION IDENTIFIED
- U UNKNOWN AND POTENTIAL ADVERSE EFFECT
- ✱ SIGNIFICANT EFFECT

AREAS OF POTENTIAL ENVIRONMENTAL EFFECTS			SITE INVESTIGATION AND PREPARATION														
			ACCESS ROADS	SITE SURVEYING	SOIL TESTING	HYDROLOGICAL TESTING	ENVIRONMENTAL SURVEY	SITE CLEARING	BURNING	EXCAVATION	DRAINAGE ALTERATION	STREAM CROSSING	EQUIPMENT	PEST CONTROL	UTILITIES	WASTE DISPOSAL AND RECOVERY	PRODUCT STORAGE
		IDENTIFICATION OF ACTIVITIES						X									
PHYSICAL-CHEMICAL EFFECTS	WATER – GROUND WATER	FLOW AND WATER TABLE ALTERATION															
		INTERACTION WITH SURFACE DRAINAGE															
		WATER QUALITY CHANGES															
	WATER – SURFACE WATER	SHORELINE AND BOTTOM ALTERATION															
		DRAINAGE FLOOD CHARACTERISTICS															
		FLOW VARIATION															
		WATER QUALITY CHANGES															
	NOISE	INTENSITY															
		DURATION															
		FREQUENCY															
							N										
	LAND	SOIL EROSION						U̸									
		FLOOD PLAIN USAGE															
		BUFFER ZONES					D	U̸									
		COMPATABILITY OF LAND USES					D	U̸									
		UNIQUE PHYSICAL FEATURES					D	U̸									
		COMPACTION AND SETTLING															
		STABILITY (SLIDES AND SLUMPS)					N	U̸									
		STRESS STRAIN (EARTHQUAKES)															
		ALTERATION OF PERMAFROST REGIME															
	ATMOSPHERE	AIR CHARACTERISTICS															
		WIND															
		ICE FOG															
		CLIMATIC CHANGES (MICRO-CLIMATE)															
ECOLOGICAL EFFECTS	SPECIES AND POPULATIONS – TERRESTRIAL	TERRESTRIAL VEGETATION					N	U̸									
		WILDLIFE (INCLUDING WATERFOWL)					N	U̸									
	SPECIES AND POPULATIONS – AQUATIC	AQUATIC FURBEARERS															
		FISH															
	HABITATS AND COMMUNITIES – TERRESTRIAL	TERRESTRIAL HABITATS					N	U̸									
		TERRESTRIAL COMMUNITIES					N	U̸									
	HABITATS AND COMMUNITIES – AQUATIC	AQUATIC HABITATS															
		AQUATIC COMMUNITIES															
AESTHETIC EFFECTS	LAND	RELIEF AND TOPOGRAPHIC CHARACTER					N	U̸									
	ATMOSPHERE	ODOUR															
		VISUAL															
		SOUNDS															
	WATER	APPEARANCE															
		ODOUR TASTE															
		LAND AND WATER INTERFACE															
	BIOTA	ANIMALS					N	U̸									
		DIVERSITY OF VEGETATION					D	U̸									
	MAN MADE OBJECTS	MAN-MADE OBJECTS															
		CONSONANCE WITH NATURE															
	COMPOSITION	COMPOSITE EFFECT					D	U̸									
		UNIQUE COMPOSITION					D	U̸									
		MOOD/ATMOSPHERE					D	U̸									

Fig. A3. Table level 2 matrix.

| CONSTRUCTION | OPERATION AND MAINTENANCE | FUTURE AND RELATED ACTIVITIES |

been made to refer the project to a Panel, then an initial environmental evaluation would not be required, as the Panel would examine all relevant issues.

Initial Environmental Evaluation (IEE)

The United Nations Environment Programme (1980) has developed a set of guidelines for the assessment of industrial activities, in which guidance is given on how to undertake an IEE. Figure A4 indicates the various steps of a general assessment procedure within which an IEE is incorporated. The initial impact identification activities commence in Step 4, in which the relationships between the project and the environment are examined for possible interactions. An IEE is then conducted in Step 5. This involves the application of tests in order to determine those elements (for example, sensitive areas) and sub-elements (for example, coastal zones) of the environment which may be subject to important impacts (see Table A8). In this questionnaire checklist, the likely effects of projects are considered by answering a series of questions related to impact types. Answers to these questions provide a simplified view of the consequences of development and consequently allow a decision to be made as to whether an EIA is required.

In the United States a decision to prepare an EIA is made on the basis of whether the proposal is 'major action significantly affecting the quality of the human environment'. This is determined by the Federal agencies which have specific criteria for identification of those typical categories of action which:

(i)   normally require an environmental impact statement (EIS);
(ii) normally do not require either an EIS or an environmental assessment (IEE); and
(iii) normally require IEEs but not necessarily EISs.

If an action were not to need an EIS, then the agency is required to involve environmental agencies, project proponents and the public, to the extent practicable, in preparing the environmental assessment which is the US term for an IEE. An environmental assessment will show whether an EIA is required or not. If no EIS is to be prepared, then a formal document based on the environmental assessment is required to establish an administrative record. This document, called a Finding of No Significant Impact (FONSI), is then available for public review for 30 days before the agency makes a final decision. FONSIs shall include a brief discussion of the needs for the proposal, of alternatives, of the environmental impacts of

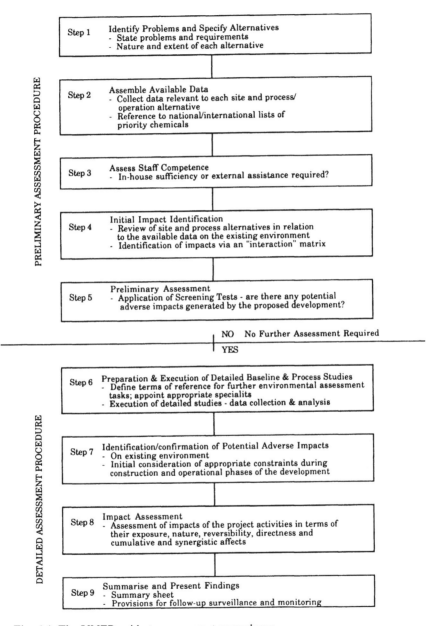

Fig. A4. The UNEP guide to assessment procedures.

Table A8
Screening table for water impacts

Sub-element	Potential impact(s)	Required information	Sources of information
Hydrological balance	Will the project alter the hydrological balance?	Extent of project; source of water, ground or other. Importance of groundwater in maintaining area rivers, streams, lakes, ponds, wells, flora and fauna	Developer Hydrologist/hydrogeologist
Groundwater regime	Will the project affect the groundwater regime, e.g. in terms of quality, quantity, depth/gradient of water table and direction of flow? Will alterations to water table depth alter structural qualities of soil? Will dewatering methods be necessary to undertake excavations?	Extent of project; source of water supply; waste disposal practices; proposed surface cover. Ground conditions: permeability, percolation, water table, location of recharge area, slope proximity to streams or other water bodies	Developer Geological maps/survey; local well-drillers; soils engineer
Drainage/channel pattern	Will the project impede the natural drainage pattern and/or induce alteration of channel form?	Existence, nature and pattern of drainage; soil characteristics	Site visit; geological maps

Category	Questions	Data needed	Source/responsible
Sedimentation	Will the project induce a major sediment influx into area water bodies?	Location of construction and cleaning activities	Developer
		Erosion potential of sites soils	Soils surveyor
		Direction of runoff flow, % slope on site	Site visit; topographical map
		Erosion and sediment control plan for site	Developer
Flooding	Will there be risk to life and materials due to flooding?	Extent of project; 100-year flood plain	Developer Geological survey
Water Quality	Does potable water supply meet established standards, WHO, etc? Will receiving waters meet established standards? Will waters be adequately accommodated and treated? Will groundwater suffer contamination by surface seepage, intrusion of saline or polluted water?	Whether existing water quality meets standards for intended usage; capacity of waste treatment plant/sewerage system to accommodate project wastes	Health/waste disposal authorities
		Water disposal plan; source of water	Developer
		Location of groundwater recharge	Hydrologist/hydrogeologist
Surface waters	Will the project impair existing surface waters through filling, dredging, water extraction or discharge; waste discharge or other detrimental practices? Will recreation or aesthetic values be endangered? Will the project affect dry weather flow characteristics?	Location of project; location of construction and clearing activities.	Developer
		Source of water supply and site of waste disposal; dams/obstructions; flow characteristics over an extended period	Civil engineer/hydrologist
		Ecological characteristics; recreation uses	Aquatic biologist, area survey

Table A9
Main considerations in the IEE guidelines for mineral projects (adapted
from EAP, 1976)

1	Overview summary	Describe the project, probable major environmental impacts, mitigating measures on significance of unmitigated environmental impacts and information deficiencies
2	Project rationale	The need, alternatives within the project and associated projects
3	Project proposal	Development concept, geological description of mineral, mining method, processing, waste disposal, supporting services etc.
4	Description of existing environment	Base-line information of the environment, climate, terrain, hydrology, biological aspects, human considerations, existing uses
5	Environmental impacts	Identification and comparison of expected environmental impacts, mitigating measures and beneficial effects
6	Major impacts and mitigating measures	Analysis of major impacts, mitigating measures and plans for surveillance and monitoring
7	Residual impacts	Unmitigated impacts should be discussed

both the action and its alternatives, and a list of agencies and persons consulted. Environmental assessments, as a result, not only determine the need for an EIS, but also give rise to improvements in those projects which do not eventually require an EIS.

As indicated in the previous section, IEEs are used in the Canadian Federal Environment Assessment and Review process. The purpose of these IEEs, is to determine whether unknown potentially adverse environmental effects are significant or whether mitigating measures can be adopted to prevent adverse effects occurring. The Canadians have developed a series of guidelines to assist the preparation of IEEs for a number of project types, e.g. mining developments (see Table A9). IEEs require deeper analysis than other screening methods and consequently more time and resources. This means that IEEs ought only to be applied as to the need for an EIA. The advantages of the IEE approach is that it allows for some public input and may also result in improvements in project design for those not subject to an EIA.

Identification of Alternatives

Few methods exist to assist the identification of alternatives, those which do, principally relate to the identification of alternative sites or routes. The overlay method, originally developed for a planning study of the town of Billerica, Massachusetts, has become established as a method for transportation alternative route selection, and has also been used to identify alternative sites. At the simplest level the key restraining factors of a development, such as, engineering factors, the existence of areas of ecological or landscape value are mapped on a transparent overlay sheet. Alternative sites or routes are then identified by placing one map on top of another. Clear areas then represent potential alternatives (see Fig. A5).

At this simple level, difficulties are encountered in dealing with more than about a dozen overlays. In addition, it is difficult to accommodate the concept of varying degrees of restraint, although to some degree, colour or tone intensity may be used. It is with the application of computerised systems that the greatest opportunities lie in relation to the identification of alternative sites or routes. The computer, using a common data set, representing the local environment under investigation can allow different restraining factors to be weighted in accordance with their importance, degrees of restraint can be easily accommodated and a final aggregate map can be produced as a numerical or shade intensity output. The computer is able not only to determine an environmental index for each alternative route, but also identify new alternative routes, which result in least environmental impacts.

Two main problems arise with this approach. First, access to data and a computer, although increasing availability of microcomputers will reduce the latter constraint. The second problem relates to the quality of the data in terms of boundary definition and heterogeneity within the grid units used to coordinate the data. For example, boundary definition in biological communities is generally difficult since gradients rather than distinct boundaries exist. Equally, a grid unit may be classified as one type of vegetation, yet contain other important types at lower densities.

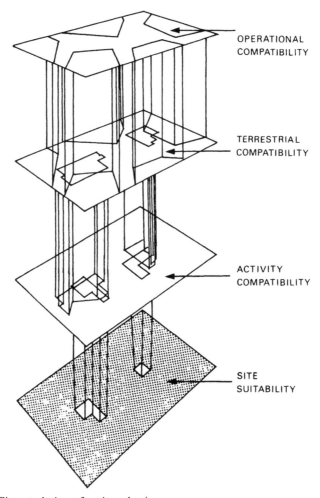

Fig. A5. Sieve technique for site selection.

Determination of Significant Issues

Two important considerations arise when considering the determination of significant issues, namely:

(i)   the range of potential significant issues; and
(ii) the identification of significant issues.

The legislative framework in which EIA operates may constrain the range of issues which may or may not be included in an EIA, for example social and economic issues may, in some countries be excluded, while in others certain aspects, such as impacts on health, may be mandatory considerations.

Scoping addresses itself to the identification of significant issues by the careful consideration of existing information relevant to the assessment, as well as the organised involvement of other agencies and consultation with the public.

Two categories of scoping have been identified. Social scoping is considered to provide the social value concerns of the public and these can be gathered by recognised methods of communication such as public meetings and structured questionnaires. Social scoping can be considered as the establishment of the terms of reference in which impacts should be considered.

Ecological scoping, on the other hand establishes the terms under which the impacts can be effectively studied, or need to be studied. The ecological scope of an EIA can be approached by focussing on four questions:

(i)   Is there reason to believe that the valued ecosystem components will be affected either directly or indirectly by the project?
(ii) Is it realistic to attempt to study the effects on the valued ecosystem components directly?
(iii) How can the effects on valued ecosystem components be studied indirectly?
(iv) Is it necessary or helpful to use indicators of impact?

Through the application of these two scoping activities significant issues may be determined. It is important, however, to remember that the term impact attaches a value to change either positive or negative and is thus related to social scoping. Change, on the other hand has no intrinsic value and it is the role of social scoping to determine the issues of social importance, while it is the role of ecological scoping to determine which changes may be predicted or measured.

As suggested, there is no method for identifying significant issues, but rather an assemblage of interactions and discussions between the public, various agencies and the project proponent.

Annex B

Environmental Policy Problems

B.1 BRIEF SUMMARY OF HEALTH IMPACT ASSESSMENT OF COAL-FIRED POWER PLANT

Primary impacts: air pollution, thermal water pollution; secondary/ tertiary impact: on drinking water quality resulting from hydrobiological disbalance caused by excessive temperatures in summertime; impact with health significance: urban air quality, drinking water quality; human exposure: people exposed to the stack plumes, people supplied by water extracted from polluted rivers.

Risk groups: people suffering from chronic respiratory diseases and people sensitive to enteric diseases; quantification of morbidity risks; increase in incidence of chronic respiratory diseases; increase in incidence of enteric diseases.

Mitigation measures: pollution control at source; siting of the plant to avoid a large population being exposed to the plume.

B.2 BRIEF SUMMARY OF HEALTH IMPACT ASSESSMENT OF WATER RESERVOIR CONSTRUCTION

Primary impact: positive — increase in water resources; negative — increase in population of mosquitoes, molluscs and other water-borne vectors or intermediate hosts of pathogens, organisms; negative/positive — modification of underground water level; other possible impact; secondary impact: decrease in water quality due to eutrophication, increase in water-borne diseases if vector infected; salination of agricultural soil if underground water level too high.

Impact with health significance: water quality if drinking water is produced from the reservoir; infectious diseases if vector/intermediate host is infected.

The salination of agricultural soil will not be considered of direct health significance, however, it can affect health through poor nutrition resulting from decrease of agricultural production.

Risk groups: people serviced by drinking water extracted from reservoir (in case of eutrophication); people living in the flying range of mosquitoes (in case of tropical diseases).

Quantification: through epidemiological data.

Mitigation measures: eutrophication control; mosquito control; resettlement of population exposed; prophylactic health care.

It is worth noting that the health impact of water development projects under a warm climate are particularly well documented.

B.3 OVERVIEW OF PROBLEMS — COMPARISON OF ALTERNATIVE ELECTRICITY PRODUCTION POLICIES

B.3.1 Electricity Production Based upon Solid Fossil Fuels: Coals, Brown-coal and/or Lignite

Due to availability of these natural resources this policy may be justified in several eastern and south-eastern European countries, and the associated problems are studied in a UNDP/World Bank assisted inter-country project on 'energy planning'. The impact analysis should not deal with individual plants but with the whole energy production cycle, wherever individual facilities are located. The cycle includes:

- *The mining stage*: If underground mining: health impact through accidents, environmental impact through subsidence (decrease of surface soil altitude which makes damming of rivers necessary).
- If open-air mining: impact on underground and surface water balance, impact on air and water quality, as well as on water reservoirs. Need to rehabilitate the holes after decommissioning.
- *The fuel-firing stage*: air pollution, and especially long-range acidic air pollution with impact on vegetation, building materials, forests, water quality and so on. Also water pollution from discharge of water used to depollute the exhaust gas, and soil pollution through disposal of ashes, residues and gypsum from SO_2 fixation on lime.

B.3.2 Electricity Production Based upon the Nuclear Cycle

Risk of major accidents; problems of long-term radioactive waste disposal; problems of reprocessing of used fuel; problem of accumulation of minute amounts of radioactive isotopes; health impact through radiation-induced cancers.

B.4 OVERVIEW OF PROBLEMS — ENVIRONMENTAL AND HEALTH IMPACT OF AGRICULTURAL POLICY

B.4.1 Putting under Cultivation, Forests or Range Lands in Tropical Countries

Environmental impact: deforestation, erosion, lateritisation, desertification.

Health impact: poor nutrition, resulting from decrease of agricultural yield following the environmental impact.

B.4.2 Intensive Agriculture in Industrial Countries

Environmental and health impact of excessive use of pesticides and fertilizers; problem of animal waste disposal from large scale feed-lots; compaction of agricultural soil under heavy mechanical equipment; health impacts through underground water contaminated by pesticides and fertilizers.

B.5 OVERVIEW OF PROBLEMS — ENVIRONMENTAL AND HEALTH IMPACT OF URBAN TRANSPORTATION POLICY

Impact of car traffic on accidents, noise, air pollution. Are clean public transport facilities available? More electric vehicles, for example. Another alternative is to decrease the demand for urban transport through town planning decisions allowing people to live close to their work. Civil engineering work built to facilitate transportation have also environmental and health impact, for example digging tunnels and trenches affect the underground water balance.

B.6 CONCLUSION

The above considerations are not answers to present-day problems but just an overview of problems to be solved. Methodology for environmental and health impact assessment does exist for most development projects and several kinds of consumer products. There is now an urgent need to work on methodological development for environmental and health assessment of alternative economic development policies and international agencies such as UNDP and the World Bank are working towards this goal.

Example of Guidelines for the Categorisation of Noxious Liquid Substances in Relation to the Aquatic Environment and Aquatic Life Hazards

Category A

Substances which are bioaccumulated and liable to produce a hazard to aquatic life or human health; or which are highly toxic to aquatic life (as expressed by a Hazard Rating 4, defined by a TLm less than 1 ppm); and additionally certain substances which are moderately toxic to aquatic life (as expressed by a Hazard Rating 3, defined by a TLm of 1 ppm or more, but less than 10 ppm) when particular weight is given to additional factors in the hazard profile or to special characteristics of the substance.

Category B

Substances which are bioaccumulated with a short retention of the order of 1 week or less; or which are liable to produce tainting of seafood; or which are moderately toxic to aquatic life (as expressed by a Hazard Rating 3, defined by a TLm of 1 ppm or more, but less than 10 ppm); and additionally certain substances which are slightly toxic to aquatic life (as expressed by a Hazard Rating 2, defined by a TLm of 10 ppm or more, but less than 100 ppm) when particular weight is given to additional factors in the hazard profile or to special characteristics of the substance.

Category C

Substances which are slightly toxic to aquatic life (as expressed by a Hazard Rating 2, defined by a TLm of 10 ppm or more, but less than 100 ppm); and additionally certain substances which are practically non-toxic to aquatic life (as expressed by a Hazard Rating 1, defined by a TLm of 100 ppm or more, but less than 1000 ppm) when particular weight is given to additional factors in the hazard profile or to special characteristics of the substance.

Category D

Substances which are practically non-toxic to aquatic life (as expressed by a Hazard Rating 1, defined by a TLm of 100 ppm or more, but less than 1000 ppm); or causing deposits blanketing the sea floor with a higher biochemical oxygen demand (BOD); or highly hazardous to human health, with an LD_{50} of less than 5 mg/kg; or produce moderate reduction of amenities because of persistency, smell or poisonous or irritant characteristics, possibly interfering with use of breaches; or moderately hazardous to human health, with an LD_{50} of 5 mg/kg or more, but less than 50 mg/kg and produce slight reduction of amenities.

Other Liquid Substances (for the purpose of Regulation 4 of this Annex)
Substances other than those categorised in Categories A, B, C and D above.

Annex D

Example of Guidelines for the Scientific Assessment of the Impact of Pollutants on the Marine Environment

These guidelines are intended to assist scientists who have been given the task of determining the potential impact of discharges into the marine environment from a particular industrial development or other human activity. Other specialists will also be assigned parallel tasks in order to assist the decision-maker. Interaction between them will assist in the decision as to whether a development or activity should proceed and under what conditions. The input of the scientist will be directed toward the development of measures to restrict impacts within acceptable limits. Some guidance is also provided on the presentation of results and for undertaking monitoring studies.

A specific procedure for investigation cannot be made concrete until or unless the discharge and design characteristics are identified. However, in all cases, a sequential procedure, similar to that outlined below, can be followed.

D.1 NATURE OF PROJECT/PROBLEM

(1) Specify the nature of the proposed development activity, discharge, etc. This should include the type of engineering activity, or process, its size and the expected characteristics (in quantity and quality terms) of the discharge.

(2) Resources requirements likely to lead to inputs to or other forms of impact on the environment should be listed, e.g. energy, water, routes of import and export of raw materials and products, ports and jetties.

(3) Interacting activities: other industry, urban growth and development, exploited commercial resources, etc. should be categorised and potential conflicts identified.

(4) The timescale of proposed developments should be charted, together with expected changes in interacting activities.

(5) At this stage, existing regulatory constraints should also be identified and negotiation with authorities initiated, so that design modifications can be incorporated early in the proceedings.

(6) Alternative options available and comparative impacts should be considered.

D.2 COLLECTION OF INFORMATION PHASE

Before embarking on the indiscriminate collection of data on the existing environment, the likely impact and relevant goals should be identified. These should reflect acceptable levels of contamination and of risk, assigned by a wider group than the scientists alone. They should also serve the needs of any models which might be used. Uncertainties may be addressed either by an explicit evaluation of the risk involved in exceeding acceptable levels or through the adoption of conservative assumptions and safety factors. An indicative list of the parameters which might have to be measured is given below. The particular circumstances will determine what is appropriate. Some data would normally be readily available, other data may need to be collected perhaps over a substantial timescale. Wherever applicable, seasonality should be taken into account. If an urgent response is called for, recourse can be made to accounts of similar problems elsewhere (see, for instance, WHO/UNEP, 1982 and WHO, 1982).

It will first be necessary to identify the potential environmental hazards. Accordingly, information should be collected on environmental behaviour and fate of raw materials, products, by-products and other associated releases such as:

- organic materials subject to rapid degradation;
- nutrients;
- persistent organic materials (including halogenated organics);
- radionuclides;
- metals and other inorganic materials;

- particulate materials;
- pathogenic microorganisms or nuisance organisms;
- energy (heated effluents or radiation);
- petroleum hydrocarbons and petrochemicals.

The literature should be reviewed for pertinent toxicity data. If data are not available, it will be necessary to initiate preliminary toxicity bioassay tests on target organisms and/or calculate dose, e.g. to man.

It will be necessary to identify sites for pre- and post-development sampling, including sites outside the expected areas of impact, and initiate investigation of, for example:

(1) Climatology: wind direction and speed, gust strength; rainfall distribution, periods of precipitation longer than 24 h; storm events.
(2) Terrestrial geology: land types/uses, topography, vegetation cover, erosion; accretion; volcanicity, seismicity; special features.
(3) Marine geology: bathymetry; sediment types and other characteristics; stability, seismicity; littoral drift (transport), erosion, accretion; special features.
(4) Marine and coastal hydrography and physico-chemical characteristics: tidal regime, currents, wave patterns, circulation; temperature, salinity, density, dissolved oxygen, alkalinity, pH; nutrients, particulate organic matter, other suspended solids.
(5) Biology: rare and endangered species; species diversity and habitats; population structure and trophic interrelationships; biomass, productivity, biochemical constituents and essential processes.
(6) Human values and uses: fishing; aquaculture, transport and communications; sand or gravel extraction, other mineral extraction in the coastal zone from the sea bed; desalination for water supply, salt and other mineral extraction from the water; waste discharges, existing and potential, domestic and industrial; archaeological, historical, aesthetic values; recreation, tourism; reserves and other special designations; human health.

D.3 ASSESSMENT OF IMPACT

From a knowledge of the nature of the contaminants and the quantities to be discharged, existing background levels, as well as biota and human uses at risk and bearing in mind existing water quality criteria or standards, the extent of impact on the receiving environment can be assessed. This will involve some, or all, of the following steps.

(1) Definition of boundary conditions
Determine environmental boundaries: based on environmental characteristics, hydrodynamics, existing uses; properties of contaminants, biogeochemical processes, kinetic parameters.

(2) Identification of targets
Consider protection of possible targets at risk: human population, habitat, food, livelihood, well-being, quality of life; plankton, intertidal species, shellfish, benthos, pelagic or demersal fish, marine birds, marine mammals, marine reptiles; egg, larval and juvenile stages; rare, endangered species or critical habitats, and the functioning of the natural environment.

(3) Pathways by which the pollutant may reach the target at risk
The following steps should be taken:

(a) Identify possible pathways through which contaminants may endanger ecosystems, human health and resources deemed to be at risk; persistent or ephemeral contaminants; food chains, bioconcentration, biomagnification.
(b) Consider the ways in which the impact or activity of the pollutant may be modified during transfer through water, sediment or biota. This will involve determination of rate of transfer, partition coefficients, rates of removal, degradability of contaminant, mean-life or mean residence time.
(c) Where practicable, use the CPA type of approach to identify the most probable route by which the pollutants involved affect the targets. It may be necessary to follow several pathways to several targets in order to establish which is the most sensitive to the impact of the development or activity. This may involve some arbitrary assumptions or the use of the probabilistic approach.

(4) Selection or derivation of standards
Where appropriate effluent standards or water quality criteria or other specific standards exist these may be used directly. Where none exist these may be derived from data from similar cases elsewhere or generated by simple toxicity testing under appropriate conditions.

(5) Calculation of environmental capacity
Environmental capacity will have to be assessed on the basis of the environmental standard selected, boundary conditions, removal processes,

etc. This will involve construction of some form of model which might range from a simple conceptual model based on mean residence time for example, to more complex ones requiring numerical or probabilistic approaches. Iterative assessment of the procedure may be advisable to refine some of the assumptions that may have been made in a first approximation.

(6) Determination of acceptable discharge rates
Based on the derived environmental capacity, an allowable input rate or quantity can be defined. Depending upon the degree of certainty with which the calculation was made, scientific prudence would normally lead to recommending that only a fraction of the input rate initially calculated should be discharged.

(7) Design and treatment options
Assess available on-site options with respect to technology, effluent and waste treatment, to establish whether they are capable of meeting the defined input rates or quantities.

(8) Overall impact assessment
After identifying the potential targets, it may be useful, as a preliminary step towards defining those which represent the key resources or critical targets to be protected, to summarise in a matrix those resources and environmental parameters which may be affected by the proposed development and/or the contaminants which may be released. To this end the probable effects can be assigned qualitative scores to describe positive ($+$), indifferent (o), negative ($-$) and double negative ($--$). In this way the environmental impact assessment matrix can serve as a primary basis to encompass many aspects involved and provides material to link scientific, social and economic facets of the case.

D.4 DECISION PHASE/PRESENTATION OF RESULTS

On the basis of the foregoing activities it will be possible for the scientist to offer advice to the regulatory agencies and/or developers as to the constraints necessary to protect the marine environment. The data and advice must be communicated in a clear and simple form to the various parties involved in the decision-making process.

The decision to go ahead with the proposed development/discharge will be made with reference to the specified constraints. If the expected discharges cannot meet the concentrations or quantities defined, alternative systems for siting, technical design, operational procedures or alternative clean-up procedures will need to be considered. The final decision will however take account of factors other than consideration of marine environmental interests alone. A final decision will rarely be clear-cut and easy.

D.5 MONITORING, VALIDATION AND REASSESSMENT PHASE

When the proposed activity has begun, or when commissioning trials are underway, the extent of impact on the receiving environment should be assessed by relating the results of baseline studies and the permitted quantities to what actually occurs, i.e. the situation should be monitored to ensure predictions were either conservative or correct. This will involve some, or all of the following steps:

(a) Establish the concentration of identified pollutant(s) in the discharge and, if variable, the frequency distribution with time.
(b) Sampling should be initiated to ascertain effects on biotic and abiotic ecosystem components.
(c) Sample identified targets within the expected exposure plume and at locations expected to be outside the influence of the plume. Observations of significant changes in population (numbers), biomass and variety of species present may be enough to establish whether the discharge is causing outside the expected natural level of variations.

If significant changes are observed, then it will be necessary:

(a) To confirm that the cause of the changes is the expected agent, or some other characteristic of the discharge alone or in conjunction with the identified effects. This may be done by further reference to toxicological studies, by field or laboratory bioassay of the target species in progressive dilutions of the discharge, and by parallel observations of
 ' no change in similar communities in areas outside the influence of the plume. The latter requirement is often difficult to achieve because of natural variability of populations, because of unchartered excursions of the plume in question, or of other discharges in the same receiving water.
(b) To establish whether such changes are likely to affect the survival, vitality, reproductive capacity or distribution of a species over a wider

area. This will entail obtaining some estimate of the local population size, and of its potential for recovery of a lost or damaged fraction or recolonisation of a damaged area.

(c) To sample food items where the target is man and food chain contamination is indicated, to establish the extent of contamination and to monitor compliance with agreed criteria for protection of the target.

If the observed effects prove to be unacceptable, methods for abatement will need to be considered. These may include limiting the period of discharge (e.g. to avoid spawning periods), limiting the volume of discharge, reducing pollutant concentrations (e.g. by pre-treatment or by changed operations), or by changed outfall design or location. In extreme cases the activity may need to cease altogether at that site, if its consequences are considered unacceptable and cannot be abated.

Even though a discharge may be designed and operated to have no unacceptable effect on the environment or on specific targets, it will be necessary to ensure continued compliance with the defined conditions. This can, in many cases, be restricted to a programme involving measurement of only a few critical parameters. Changes in 'acceptability', e.g. in legislative restrictions or in public perception, must also be taken into account.

It may be necessary to reassess the allowed discharge limits if monitoring reveals any of the following:

(a) changes in operational procedures or levels of activity (and effects of changes assessed and monitored), as well as other, independent, changes influencing the same receiving waters;
(b) changes in biological or abiotic status of the receiving water;
(c) changes in the status of the defined target(s).

In any event, the position should be reassessed in the light of new scientific data, new treatment technologies, refinements in earlier assumptions, confirmation/refutation of initial impact assessment and revision in regulatory procedures.

Contingency plans should be formulated; likely malfunctions and other emergencies should be identified and plans developed to minimise impacts.

List of Authors and Papers Used on Handbook

Affiliation at time of presentation of paper
Chapter 1
- The World Health Organization Interest in EIA: Eric Giroult, WHO Regional Office for Europe, Copenhagen, 1988.
- Aims and Objectives of EIA: Brian Clark, CEMP, Aberdeen, 1987.
- The Health and Safety Component of EIA: Brian Clark, CEMP, Aberdeen, 1987.

Chapter 2
- Introduction to EIA Methods: Characteristics and Uses of Matrices and Networks: Ron Bisset, CEMP, Aberdeen, 1988.
- Environmental Impact Assessment: Process, Methods and Uncertainty, Ron Bisset, CEMP, Aberdeen, 1987.
- Screening and Scoping Methods: Paul Tomlinson, CEMP, Aberdeen, 1988.

Chapter 3
- Principles and Techniques of Epidemiology in Environmental Health: Lloyd & Williams, 1987.
- Introduction to Occupational Hygiene: Bradley and Dodgson, Institute of Occupational Medicine, Edinburgh, 1987.
- Principles of Toxicology: M.H. Draper, 1987.
- Methods of Environmental Health Impact Assessment: J. Martin, Cobinam Resource Consultants, Oxford, 1988.

Chapter 4

- Chemical Risk Assessment: Dinko Kello, WHO Regional Office for Europe, Copenhagen, 1987.
- Chemicals in the Environment: Giovanni Zapponi, Istituto Superiore di Senita, Rome, 1988.
- Role of the International Programme on Chemical Safety: Professor M. Mercier, IPCS, Geneva, 1987.
- Risk Assessment: Alan Ryder, CEMP, Aberdeen, 1988.
- EHIA for Consumer Products, Development Projects and Development Policies: Stern and Giroult, WHO Regional Office for Europe, Copenhagen, 1987.
- Qualitative and Quantitative Toxicological Assessment in EHIA: Anna Rita Bucchi Istituto Superiare di Sanita, Rome, 1987.

Chapter 5

- Health Hazard Factor Identification: Giovanni Zapponi, Istituto Superiore di Sanita, Rome, 1987.
- Methods for the Health Component of Industrial Development Projects: Giovanni Zapponi, Istituto Superiore di Sanita, Rome, 1988.

Chapter 6

- Presentation of Health Impact Predictions: Alan Ryder, CEMP, Aberdeen, 1987.
- Introduction to Methods for Considerations of the Health Component of EIA: Margaret Mogford, CEMP, Aberdeen, 1987.
- The Relevance of EIA to the Work of the International Maritime Organisation: David Edwards, IMO, London, 1988.
- The Role of Auditing and Monitoring in EIA: Matt Davies, CEMP, Aberdeen, 1988.

Chapter 7

- Why and How to Strengthen Human Health Considerations in EIA: Eric Giroult, WHO Regional Office for Europe, Copenhagen, 1988.

Chapter 8

- Forecasting the Vector-Borne Disease Implications of Irrigation Projects: A Case Study in the Republic of Zambia: Dr M. Birley, Department of Medical Entomology, Liverpool School of Tropical Medicine, London.

- Saguling Dam/Reservoir (Indonesia) and Nam Pong Project (Thailand): Ron Bisset, CEMP, Aberdeen, 1988.
- Health Aspects in EIA of Two Industrialised Areas in Poland: Anna Starzewska, Environmental Abatement Centre Katowice, Poland, 1988.
- Industrial Development: Iron Smelting Plants in Brazil: Ricardo A. Braun, 1988.
- Industrial Development: The Seveso Accident: Giovanni Zapponi, Istituto Superiore di Sanita, Rome, 1987.
- Health Impacts of Water Development in Turkey: Utku Unsal, Ministry of Health Ankara, Turkey, 1987.

Index

List of Tables

List of Figures

Printed and bound by CPI Group (UK) Ltd, Croydon, CR0 4YY

23/10/2024

01777667-0004